Vortical Solutions of the Conical Euler Equations

by Kenneth G. Powell

Notes on Numerical Fluid Mechanics (NNFM) Volume 28

Series Editors: Ernst Heinrich Hirschel, München
Kozo Fujii, Tokyo
Keith William Morton, Oxford
Earll M. Murman, M.I.T., Cambridge
Maurizio Pandolfi, Torino
Arthur Rizzi, Stockholm
Bernard Roux, Marseille

(Adresses of the Editors: see last page)

Volume 6 Numerical Methods in Laminar Flame Propagation (N. Peters/J. Warnatz, Eds.)
Volume 7 Proceedings of the Fifth GAMM-Conference on Numerical Methods in Fluid Mechanics (M. Pandolfi/R. Piva, Eds.)
Volume 8 Vectorization of Computer Programs with Applications to Computational Fluid Dynamics (W. Gentzsch)
Volume 9 Analysis of Laminar Flow over a Backward Facing Step (K. Morgan/J. Periaux/F. Thomasset, Eds.)
Volume 10 Efficient Solutions of Elliptic Systems (W. Hackbusch, Ed.)
Volume 11 Advances in Multi-Grid Methods (D. Braess/W. Hackbusch/U. Trottenberg, Eds.)
Volume 12 The Efficient Use of Vector Computers with Emphasis on Computational Fluid Dynamics (W. Schönauer/W. Gentzsch, Eds.)
Volume 13 Proceedings of the Sixth GAMM-Conference on Numerical Methods in Fluid Mechanics (D. Rues/W. Kordulla, Eds.) (out of print)
Volume 14 Finite Approximations in Fluid Mechanics (E. H. Hirschel, Ed.)
Volume 15 Direct and Large Eddy Simulation of Turbulence (U. Schumann/R. Friedrich, Eds.)
Volume 16 Numerical Techniques in Continuum Mechanics (W. Hackbusch/K. Witsch, Eds.)
Volume 17 Research in Numerical Fluid Dynamics (P. Wesseling, Ed.)
Volume 18 Numerical Simulation of Compressible Navier-Stokes Flows (M. O. Bristeau/R. Glowinski/J. Periaux/H. Viviand, Eds.)
Volume 19 Three-Dimensional Turbulent Boundary Layers – Calculations and Experiments (B. van den Berg/D. A. Humphreys/E. Krause/J. P. F. Lindhout)
Volume 20 Proceedings of the Seventh GAMM-Conference on Numerical Methods in Fluid Mechanics (M. Deville, Ed.)
Volume 21 Panel Methods in Fluid Mechanics with Emphasis on Aerodynamics (J. Ballmann/R. Eppler/W. Hackbusch, Eds.)
Volume 22 Numerical Simulation of the Transonic DFVLR-F5 Wing Experiment (W. Kordulla, Ed.)
Volume 23 Robust Multi-Grid Methods (W. Hackbusch, Ed.)
Volume 24 Nonlinear Hyperbolic Equations – Theory, Computation Methods, and Applications (J. Ballmann/R. Jeltsch, Eds.)
Volume 25 Finite Approximations in Fluid Mechanics II (E. H. Hirschel, Ed.)
Volume 26 Numerical Solution of Compressible Euler Flows (A. Dervieux/B. Van Leer/J. Periaux/A. Rizzi, Eds.)
Volume 27 Numerical Simulation of Oscillatory Convection in Low-Pr Fluids (B. Roux, Ed.)
Volume 28 Vortical Solutions of the Conical Euler Equations (K. G. Powell)
Volume 29 Proceedings of the Eighth GAMM-Conference on Numerical Methods in Fluid Mechanics (P. Wesseling, Ed.)
Volume 30 Numerical Treatment of the Navier-Stokes Equations (W. Hackbusch/R. Rannacher, Eds.)

Vortical Solutions of the Conical Euler Equations

by Kenneth Grant Powell

Manuscripts should have well over 100 pages. As they will be reproduced photomechanically they should be typed with utmost care on special stationary which will be supplied on request. In print, the size will be reduced linearly to approximately 75 per cent. Figures and diagramms should be lettered accordingly so as to produce letters not smaller than 2 mm in print. The same is valid for handwritten formulae. Manuscripts (in English) or proposals should be sent to the general editor Prof. Dr. E. H. Hirschel, Herzog-Heinrich-Weg 6, D-8011 Zorneding.

Vieweg is a subsidiary company of the Bertelsmann Publishing Group International.

All rights reserved
Copyright © 1987 Massachusetts Institute of Technology
Published by Friedr. Vieweg & Sohn Verlagsgesellschaft mbH, Braunschweig 1990

No part of this publication may be reproduced, stored in a retrieval system or transmitted, mechanical, photocopying or otherwise, without prior permission of the copyright holder.

Produced by W. Langelüddecke, Braunschweig
Printed in the Federal Republic of Germany

ISSN 0179-9614

ISBN 3-528-07627-5

Foreword

This book contains a study of supersonic flows past delta wings. Computational, analytical and experimental results are presented, with the goal of reaching a better understanding of leading-edge vortex flows. An algorithm for the numerical solution of the governing equations of inviscid compressible flow is developed. The method allows regions of local refinement in the mesh, for high resolution solutions at low computational cost. Results from the scheme are presented, along with questions that arise from the results. The primary questions that arise pertain to the highly localized regions of total pressure loss that occur in the vicinity of the vortices. A model for these losses that is consistent with their behavior is postulated, based on a similarity solution of the axisymmetric Navier-Stokes equations. Finally, results from the solution scheme are compared with experimental data, to gain further insight into the physics of these flows.

The work presented in this book is a result of my doctoral studies at the Massachusetts Institute of Technology. The book is meant primarily as a research monograph on the topic of the computation of leading-edge vortex flows.

"Independent research," the catch-phrase often associated with doctoral programs, conjures up the image of a lonely graduate student, slaving away in some dingy garret. Happily this image bears little resemblance to the way that the work presented here was carried out. I had help from a number of quarters, and my office at M.I.T. was quite well-lit and fairly pleasant, if a little over-air-conditioned.

First, I'd like to thank Professor Earll Murman for — well, for everything. He showed an uncanny knack of knowing when I needed a gentle nudge in the right direction, when I needed encouragement and when I just needed a sympathetic ear. I'd also like to thank him for always treating me as an equal, a status I hope someday to deserve.

I'd also like to thank my doctoral committee members and thesis readers: Professor Eugene Covert, who first kindled my interest in fluid dynamics; Professor Mårten Landahl, who always seemed to be able to find a new and different approach to a problem; Professor James McCune, who helped provide ideas and enthusiasm in our vortex flow working group; and Dr. Bernhard Müller, who read this document more carefully than I would have thought humanly possible. Thanks also to Professor Judson Baron, for getting me started in vortex flows, and to Professor Michael Giles, for help along the way.

I owe thanks to all of the students of the Computational Fluid Dynamics lab, but I would especially like to thank my "contemporaries," Dr. John Dannenhoffer and Dr. Rich Shapiro, for all their help. I'd also like to thank Björn Krouthén for being such a terrific office-mate, and for expanding my knowledge of the off-color portions of the Scandinavian languages. Thanks, also, to Dr. Tom Roberts, for laughing at my more obscure jokes. Bob Bruen and Teva Regule kept the computers running, and were awfully good company, too. Ellen Mandigo proofread this manuscript, and kept me posted on the relevés and pliés of the Boston Ballet.

Technical help also came from outside MIT. Thanks to Dave Miller and Rick Wood for experimental data and a stimulating exchange of ideas. Thanks to Dr. Frank Marconi for writing the paper that first got me interested in this problem, and for his suggestions along the way. Thanks to Jim Thomas and Rick Newsome for their help, particularly while I was at NASA Langley. And thanks to Professors Bram van Leer and Tom Adamson here at the University of Michigan — the thought of working with them gave me incentive to finish my degree.

On a more personal note, thanks are due to some very special friends: to Max and Alta (and Erik) Blosser for putting up with me for a summer, and for being such wonderful hosts; to Ed Kearns and Heidi Harvey and Rich and Crystal Schaaf for their full-time friendship and patience; to Denise Cormier for being a terrific friend; and to Sanne Krummel, my light-hearted and light-footed friend, for the support, inspiration and joy she provides me.

Finally, to my family — to my parents, without whom life itself would be impossible, and to my brothers, John and Tom: for support of all kinds; for always telling me I work too hard (even when I don't); for whisking me off to exotic places to help me forget work once in a while; and for teaching me how important a sense of humor is.

Kenneth G. Powell

Contents

1 Introduction   1
    1.1 Analysis of Vortex Flows 4
         1.1.1 Potential Methods 4
         1.1.2 Euler and Navier-Stokes Methods 5
    1.2 Present Research 7

2 Governing Equations   9
    2.1 Euler Equations 9
    2.2 Non-Dimensionalization 12
    2.3 Boundary Conditions 13
    2.4 Conical Self-Similarity 13
    2.5 Topological Considerations 15
    2.6 Summary 17

3 Solution Procedure   19
    3.1 Spatial Discretization 19
         3.1.1 Grid Generation 19
         3.1.2 Finite Volume Formulation 26
         3.1.3 Flux Integration 27
    3.2 Artificial Viscosity 27
    3.3 Temporal Discretization 30
    3.4 Boundary Conditions 31
         3.4.1 Solid Wall Condition 31
         3.4.2 Far-Field Condition 32

		3.4.3	Kutta Condition	33
		3.4.4	Symmetry Plane Condition	33
		3.4.5	Embedding Interfaces	33
	3.5		Artificial Viscosity Boundary Conditions	35
	3.6		Data Structure	39
	3.7		Summary	40

4 Basic Characteristics of Solutions 43

  4.1 Topological Aspects of the Solutions 43
  4.2 Pressure Plots 47
  4.3 Mach Number and Density 51
  4.4 Vorticity 55
  4.5 Grid Resolution Effects 55
  4.6 Convergence 60
  4.7 Summary 61

5 Total Pressure Losses — A Numerical Study 63

  5.1 Effects of Computational Parameters 64
  5.2 Effects of Physical Parameters 78
  5.3 Comparison of Loss with Experiment 79
  5.4 Artificial Viscosity Level in the Vortex 88
  5.5 Lossless Solutions to the Euler Equations 95
  5.6 Summary 100

6 Total Pressure Losses — A Theoretical Model 101

  6.1 Model for the Feeding Sheet 101
  6.2 Burgers' Vortex 104
  6.3 Model for Total Pressure Loss 106
  6.4 Summary 116

7 Comparison of Computations and Experiments 119

  7.1 Experimental Setup and Models 119
  7.2 Flat plate cases 120
   7.2.1 Case 1 — $M_\infty = 1.7$, $\alpha = 12°$, $\Lambda = 75°$ 121

		7.2.2	Case 2 — $M_\infty = 2.0$, $\alpha = 20°$, $\Lambda = 60°$ 127

       7.2.3 Case 3 — $M = 2.8$, $\alpha = 20°$, $\Lambda = 75°$ 138
  7.3 Vortex flap cases 148
       7.3.1 Case 1 — $M_\infty = 2.4$, $\alpha = 4°$, $\delta = 10°$ 148
       7.3.2 Case 2 — $M_\infty = 2.4$, $\alpha = 12°$, $\delta = 10°$ 152
  7.4 Asymmetric cases 162
       7.4.1 Case 1 — $M_\infty = 1.7$, $\alpha = 12°$, $\beta = 8°$ 162
       7.4.2 Case 2 — $M_\infty = 2.8$, $\alpha = 12°$, $\beta = 8°$ 166
  7.5 Summary 175

8 Conclusions   183

  8.1 Algorithm Development 183
  8.2 Characteristics of the Solutions 184
  8.3 Comparison of Numerical and Physical Results 185
  8.4 Recommendations for Further Research 185

Bibliography   187

A Stability Analysis   193

B Cross-flow streamline integration   203

C Computer Code   205

   List of Symbols   281

   Subject Index   283

Chapter 1

Introduction

Vortices, the "sinews and muscles of fluid flow" [29], have attracted researchers in fluid mechanics from Leonardo da Vinci, if not earlier, onwards. Leading-edge vortex flows in particular are of great interest to aerodynamicists, predominantly for three reasons. The first is that lee-side vortices, when symmetric and stable, can lead to beneficial high-lift situations. The second is that lee-side vortices, when asymmetric or unstable, can lead to disaster, in the form of spin or stall. The third reason is aesthetic in nature. Very simple geometries, such as a flat plate delta wing at angle of attack, can lead to extremely intricate and intriguing flow topologies. High-speed, high angle of attack flight, in particular, produces a rich variety of shock-vortex interaction patterns.

The basic problem examined in this book is that of a flat plate delta wing in a supersonic free-stream. The best known flow that occurs in this problem is one with separation at the leading-edges of the wing. This results in large primary vortices on the leeward side of the wing, with smaller, viscous-induced secondary vortices beneath them (See Figure 1.1). This is by no means the only possible flow topology, however. Stanbrook and Squire [59] postulated four main types of flows for highly swept wings:

1. Flow separating at the leading edge, rolling up into a vortex;

2. Flow attached at the leading-edge — no shock waves inboard;

3. Flow attached at the leading-edge — shock wave inboard, no separation;

4. Flow attached at the leading-edge — shock wave inboard, causing separation there.

They empirically classified these flows, using the component of the Mach number normal

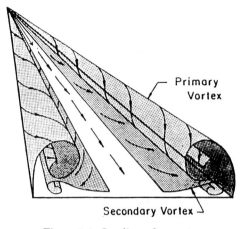

Figure 1.1: Leading-edge vortex

to the leading-edge, M_N, and the angle of attack normal to the leading-edge, α_N, as the correlation factors. Their data suggested a boundary near $M_N = 1$: to the left of the boundary ($M_N < 1$) the leading-edge is swept behind the Mach cone and the flow is separated at the leading-edge; to the right ($M_N > 1$), the flow is attached, with a Prandtl-Meyer expansion at the leading-edge.

Miller and Wood [36] carried the classification further, adding the following cases:

5. Flow separating at the leading edge, resulting in a vortex with a shock above it;

6. Separation bubble at the leading-edge — no shock waves on the bubble;

7. Separation bubble at the leading-edge — shock wave above the bubble.

They also classified the flows using M_N and α_N, producing the chart of flow regimes diagrammed in Figure 1.2. Vorropoulos and Wendt [67] postulated the possibility of leading-edge vortex flow with a reverse cross-flow shock under the vortex (See Figure 1.3), based on evidence from their laser-doppler results. This flow regime was at low normal Mach number and angle of attack.

It is exactly this richness of flow phenomena that makes this problem such an exciting one. The possibility of finding new complex flow topologies for this geometrically simple problem remains, both for experimentalists and theoreticians.

In the next section of this chapter, a brief survey of the methods used to model these flows is given. The final section of this chapter describes the goals of the research presented in this book, and provides an outline for the remainder of the book.

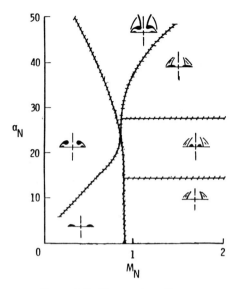

Figure 1.2: Flow regime diagram

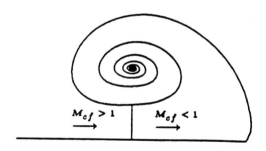

Figure 1.3: Reverse cross-flow shock under vortex

1.1 Analysis of Vortex Flows

The methods of analyzing leading-edge vortex flows are of two basic types. The first consists of methods which model the vortex in some approximate way, based on *a priori* knowledge of the topology of the flow. The best known method of this type is the Polhamus suction analogy [47], based on the leading-edge suction force in classical lifting surface theory. The *ad hoc* assumption of the Polhamus analogy is that the normal force due to the vortex is equal to the suction force that would occur in attached flow, rotated 90°. The second type, and the one primarily considered here, consists of methods which model the vortex using a set of governing equations for the flow — the potential equation, the Euler equations or the Navier-Stokes equations. Fundamental issues concerning these methods will be considered in the next two sections.

1.1.1 Potential Methods

Since these methods are based on the potential equation, vorticity must be introduced into the flow as a singularity of some type. One method is the point vortex method, introduced by Rosenhead [56] and extended and popularized by Moore [38], in which the (scalar) vorticity field is given as a sum of Dirac delta functions

$$\omega = \sum_{i=1}^{N} \Gamma_i \delta\left(\mathbf{x} - \mathbf{x_i}\left(\mathbf{t}\right)\right), \quad (1.1)$$

where $\mathbf{x_i}$ is the location of the ith vortex, and Γ_i is its circulation. The evolution of the vortices in time is obtained by a Lagrangian integration of the vorticity transport equation

$$\frac{D\omega}{Dt} = 0, \quad (1.2)$$

where the velocity field is derived from the vorticity *via* the Biot-Savart relation. This method has the unlucky feature of being ill-posed for the initial value problem of the roll-up of a vortex sheet, tending to become chaotic in the spiral region. The replacement of the point vortices with vortices of finite core size has been presented by Chorin and Bernard [11] as a method of stabilizing the sheet. This is done at the expense of consistency, however, as it is difficult to assess the errors introduced by this modification [4].

1.1. ANALYSIS OF VORTEX FLOWS

Other similar methods include contour dynamics, in which the initial vorticity is assumed to be regionwise constant and the boundaries of these regions are tracked in a Lagrangian manner [68], and cloud-in-cell techniques, which are hybrid Eulerian-Lagrangian techniques in which an Eulerian mesh is used to solve the Poisson equation for the velocity field [4,42].

Vortex sheet methods began with Brown and Michael [6] in the modeling of the leading-edge vortex as a concentrated line vortex connected to the wing by a cut, such that the force on the cut balances the Joukowski force on the line vortex. Mangler and Smith [32] later replaced the cut with a curved vortex sheet. Panel methods were subsequently used, in which the vortex is modeled by a sheet of discrete panels, across which a zero mass-flux and pressure jump condition are enforced [22,25].

Since the numerical handling of a spiral vortex sheet containing an infinite number of turns would present some difficulty, it is necessary to replace the inner portion of the spiral with some type of core model. Hall [19] and Stewartson and Hall [61] developed a core model based on an asymptotic expansion with a conical inviscid outer solution and a non-conical viscous inner solution. Luckring [31] matched the solution of Stewartson and Hall to an outer spiralling vortex sheet. Brown [7] carried out an analysis for a compressible, inviscid vortex that followed Hall's outer solution analysis. Mangler and Weber [33] found an asymptotic expansion for the core which contained Hall's solution as its leading term. Guiraud and Zeytounian [18] took advantage of the double-scale form inherent in the model of Mangler and Weber to derive a two-scale analysis based on the slenderness of the vortex and the "closeness" of the turns of the spiral. Brown and Mangler [8] extended the model of Mangler and Weber to account for the effects of compressibility. These core models are made necessary by the inability of the potential equation to model distributed vorticity. For a model which "captures" distributed vorticity, one must turn to the Euler or Navier-Stokes equations.

1.1.2 Euler and Navier-Stokes Methods

Euler and Navier-Stokes methods consist of discretizing the governing equations on a grid to solve them in an approximate manner. The Euler equations are attractive as a solution method for vortex flows because they permit solutions with distributed vorticity, and are less expensive to solve than the Navier-Stokes equations. They model the primary vortex and shock waves, but cannot account for the viscous effects in the boundary layer or the secondary separation. The Navier-Stokes equations are chosen to model flows in which viscous effects play a large role. In principle, they can model all the features of a leading-edge vortex flow. In practice, however, the accuracy of the solutions is severely limited by speed and memory constraints of available computers and difficulties in turbulence modeling [44].

Solutions of the Euler equations using centered-difference conservative schemes have been carried out by Rizzi [54], Rizzi and Eriksson [55], Murman *et al* [41], Newsome [43], Kandil and Chuang [27] and Arlinger [3] among others. Upwind-difference conservative schemes have been used by Newsome and Thomas [45] and Chakravarthy and Ota [10]. Characteristic-based non-conservative schemes have been used by Marconi [34]. All of these give reasonable estimates for the pressure coefficient on the wing, although they tend to overpredict the suction peak due to the vortex. There is a fundamental difference between the λ-scheme solutions of Marconi and the conservative scheme solutions of the other authors, however. In the λ-scheme, the only entropy production is through shocks that are fitted into the flow-field. In the conservative schemes, large losses are consistently seen in the region of the vortex and the feeding sheet. While it is not immediately apparent why discrete solutions to the conservation form of the Euler equations exhibit these losses, comparison of the losses with those seen experimentally and independence of computational parameters such as mesh spacing and artificial viscosity of the level of loss suggest an underlying physical principle.

Solutions of the Navier-Stokes equations have been reported by Vigneron *et al* [66], Rizetta and Shang [53], Fujii and Kutler [16], Newsome and Thomas [45], Thomas *et al* [63] and Müller and Rizzi [39] among others. The solutions tend to show better agreement with experiment than Euler solutions, although insufficient resolution marred some of the results.

The Navier-Stokes and Euler equations models tend to give approximately the same solution in the region of the primary vortex [35]. The similarity of these solutions raises questions as to whether Euler methods are truly solving the Euler equations and the Navier-Stokes methods are truly solving the Navier-Stokes equations, or whether truncation error and added artificial viscosity are modifying the equations beyond recognition. This points out a need to study the way that vortex cores are modelled by Euler and Navier-Stokes methods.

1.2 Present Research

The present research has three primary goals. They are:

1. To develop an efficient algorithm for modeling leading-edge vortex flows;

2. To understand the way in which the vortex is modeled by the scheme;

3. To gain insight into the physics of leading-edge vortex flows, using a synergistic blend of computational and experimental results.

For the first goal, a conical Euler equation model is chosen. The reasons for the use of the conical self-similarity assumption are:

- The fact that the Euler equations permit a conical solution for supersonic flow past the geometries considered;

- Experimental evidence of the conical nature of these flows;

- The great reduction in computational cost relative to three-dimensional solutions.

The Euler equations and conical self-similarity assumption are covered in Chapter 2.

The solution method chosen for the conical Euler equation model is a cell-vertex scheme that allows for multiple levels of embedded grids. The embedded grid scheme is chosen to allow good resolution in the leading-edge vortex region of the flow without the necessity of many grid points in the rest of the flow. The cell-vertex scheme is chosen for its second-order accuracy, even at embedding interfaces and regions of large grid stretching or skewness. The solution scheme is outlined in Chapter 3.

The second goal of the research presented herein is to understand the way in which the scheme models the flow. In Chapter 4, the basic characteristics of the solutions are examined. In Chapter 5, the total pressure losses that occur in the vortex and its feeding sheet are studied. The effects of various computational and physical parameters on the loss are examined. In Chapter 6, the way that the method models the feeding sheet and the vortex core is discussed. The way that the vortex sheet is modeled is explained by looking at the form that the discrete Euler equations take in a vortex sheet. The way that the core is modeled is explained by a new similarity solution to the Navier-Stokes equations for incompressible, axisymmetric conical flow.

The third goal of the research is to gain physical insight into these flows. Chapter 7 presents numerical and experimental results for several cases, including the effects of yaw and vortex flaps. The experimental results give insight into phenomena that the computations cannot model and serve to produce confidence in the numerical results. The computations allow one to examine flow parameters that are difficult to measure experimentally. This synergistic approach to the study of these flows provides a great deal of insight into what is happening.

Chapter 2

Governing Equations

In this chapter, the governing equations used in this book are derived, and appropriate choices of non-dimensionalization and boundary conditions are described. The conical self-similarity assumption is defined, and its implications on the topology of the flow are explained.

The governing equations used for this study are the conical Euler equations for an ideal gas. The assumptions necessary for these equations to hold are:

- The fluid is a homogeneous, non-conducting ideal gas;
- The Reynolds number, $\rho U_\infty L/\mu$, is infinite;
- The Peclet number, $\rho U_\infty C_p L/k$, is infinite;
- The flow field is conically self-similar.

These assumptions strictly limit the class of problems to supersonic, inviscid flow over a conically self-similar geometry. The Reynolds number and Peclet number will not truly be infinite in physical situations, but must be large enough that the viscous and thermal boundary layers may be ignored. Although only steady solutions of the conical Euler equations are considered here, these solutions are achieved by solving the unsteady Euler equations in a time-asymptotic fashion. Therefore the derivations below are for the unsteady Euler and conical Euler equations.

2.1 Euler Equations

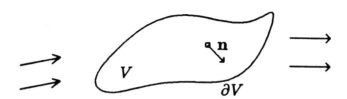

Figure 2.1: Control volume for conservation laws

The Euler equations are a set of partial differential equations describing the state of a homogeneous, non-conducting fluid with infinite Reynolds number, infinite Peclet number, and no body force. They are derived below from the integral form of the conservation laws.

The basic laws governing the motion of a homogeneous non-conducting fluid are the conservation of mass, linear momentum and energy. The conservation of mass may be written in integral form as

$$\iiint_V \frac{\partial \rho}{\partial t} d^3x + \iint_{\partial V} \rho u_i n_i d^2x = 0. \tag{2.1}$$

This states that the time rate of change of density within a fixed volume V is balanced by the mass flux through the boundary ∂V. In the above, d^3x is a volume element, d^2x is a surface element, and n_i is a component of the surface normal (See Figure 2.1). Gauss' theorem may be used to write

$$\iiint_V \left[\frac{\partial \rho}{\partial t} + \frac{\partial}{\partial x_i}(\rho u_i) \right] d^3x = 0, \tag{2.2}$$

which yields the differential form of the mass conservation equation,

$$\frac{\partial \rho}{\partial t} + \frac{\partial}{\partial x_i}(\rho u_i) = 0. \tag{2.3}$$

The differential form holds everywhere but in the region of a discontinuity in the flow. At a discontinuity, only the integral forms Eqs. (2.1) and (2.2) hold.

The conservation of momentum may be written in integral form as

$$\iiint_V \frac{\partial}{\partial t}(\rho u_i) d^3x + \iint_{\partial V} \rho u_i u_j n_j d^2x = \iiint_V \rho F_i d^3x + \iint_{\partial V} \sigma_{ij} n_j d^2x. \tag{2.4}$$

2.1. EULER EQUATIONS

This states that the rate of change of momentum of the fluid within a volume V is balanced by the momentum flux into the volume, the volume force F_i and the surface stress tensor σ_{ij} acting on the fluid. Gauss' theorem may be used to obtain the differential form,

$$\frac{\partial}{\partial t}(\rho u_i) + \frac{\partial}{\partial x_j}(\rho u_i u_j) = \rho F_i + \frac{\partial \sigma_{ij}}{\partial x_j}. \tag{2.5}$$

Again, this form is valid only in continuous regions of the flow.

The conservation of energy may be written in integral form as

$$\iiint_V \frac{\partial}{\partial t}(\rho E)\, d^3x + \iint_{\partial V} \rho E u_i n_i\, d^2x = \iint_{\partial V} k \frac{\partial T}{\partial x_i} n_i\, d^2x + \iint_{\partial V} u_i \sigma_{ij} n_j\, d^2x, \tag{2.6}$$

where E is the mass-specific energy, T is the static temperature and k is the coefficient of thermal conductivity. This states that the rate of change of energy within a volume V is balanced by the heat transfer due to conduction and the work done by the surface forces acting on the volume. The differential form of the energy equation, valid in continuous regions, is

$$\frac{\partial}{\partial t}(\rho E) + \frac{\partial}{\partial x_i}(\rho E u_i) = \frac{\partial}{\partial x_i}\left(k \frac{\partial T}{\partial x_i}\right) + \frac{\partial}{\partial x_j}(u_i \sigma_{ij}). \tag{2.7}$$

The stress tensor in the equations of momentum and energy may be expressed as a sum of isotropic and non-isotropic parts,

$$\sigma_{ij} = -p\, \delta_{ij} + d_{ij}, \tag{2.8}$$

where p is the static pressure of the fluid and d_{ij} is the non-isotropic portion of the stress. To obtain the equations of motion for an inviscid fluid, it is assumed that the stress tensor is isotropic, i.e.,

$$\sigma_{ij} = -p\, \delta_{ij}. \tag{2.9}$$

This is true in the case of infinite Reynolds number. With this assumption, and the assumption that the body force F_i is negligible, Equation (2.5) becomes

$$\frac{\partial}{\partial t}(\rho u_i) + \frac{\partial}{\partial x_j}(\rho u_i u_j + p\, \delta_{ij}) = 0. \tag{2.10}$$

With the assumption of isotropic stress, and ignoring the heat transfer terms (which is equivalent to assuming infinite Peclet number), Equation (2.7) becomes

$$\frac{\partial}{\partial t}(\rho E) + \frac{\partial}{\partial x_i}([\rho E + p]\, u_i) = 0. \tag{2.11}$$

The mass conservation equation, Equation (2.3), along with Equations (2.10) and (2.11), may be written in vector form for an (x, y, z) coordinate system as

$$\frac{\partial}{\partial t}\begin{bmatrix} \rho \\ \rho u \\ \rho v \\ \rho w \\ \rho E \end{bmatrix} + \frac{\partial}{\partial x}\begin{bmatrix} \rho u \\ \rho u^2 + p \\ \rho uv \\ \rho uw \\ \rho u h_0 \end{bmatrix} + \frac{\partial}{\partial y}\begin{bmatrix} \rho v \\ \rho uv \\ \rho v^2 + p \\ \rho vw \\ \rho v h_0 \end{bmatrix} + \frac{\partial}{\partial z}\begin{bmatrix} \rho w \\ \rho uw \\ \rho vw \\ \rho w^2 + p \\ \rho w h_0 \end{bmatrix} = 0 , \qquad (2.12)$$

where h_0 is the stagnation enthalpy, defined as

$$h_0 = E + \frac{p}{\rho}. \qquad (2.13)$$

These are commonly referred to as the three-dimensional Euler equations.[1] They may be expressed in terms of a state vector $\mathbf{U}$ and flux vectors $\mathbf{F}$, $\mathbf{G}$ and $\mathbf{H}$ as

$$\frac{\partial \mathbf{U}}{\partial t} + \frac{\partial \mathbf{F}}{\partial x} + \frac{\partial \mathbf{G}}{\partial y} + \frac{\partial \mathbf{H}}{\partial z} = 0. \qquad (2.14)$$

An equation of state is necessary to close the set of equations. The relation between pressure, density, energy and velocity of an ideal gas,

$$p = (\gamma - 1)\rho \left[E - \frac{u^2 + v^2 + w^2}{2} \right], \qquad (2.15)$$

is used here.

2.2 Non-Dimensionalization

It is convenient to non-dimensionalize the Euler equations for solving them numerically. The non-dimensionalization parameters chosen are the freestream density, ρ_∞, the free-stream speed of sound, c_∞, and a characteristic length scale, L. The non-dimensionalizations are given in Table 2.1. The Euler equations and the equation of state are invariant to this non-dimensionalization. Under this transformation, the non-dimensional free-stream state vector becomes

$$\begin{bmatrix} \rho \\ \rho u \\ \rho v \\ \rho w \\ \rho E \end{bmatrix}_\infty \longrightarrow \begin{bmatrix} 1 \\ M_\infty \cos\alpha \cos\beta \\ M_\infty \sin\beta \\ M_\infty \sin\alpha \cos\beta \\ \frac{1}{\gamma(\gamma-1)} + M_\infty^2/2 \end{bmatrix}, \qquad (2.16)$$

where α and β are the angles of attack and yaw, respectively.

---

[1]In classical fluid mechanics, the three momentum equations in non-conservation form are referred to as the Euler equations.

2.3. BOUNDARY CONDITIONS

Table 2.1: Non-dimensionalization factors

Variable	Factor
ρ	ρ_∞
u,v,w	c_∞
E, h_0	c_∞^2
p	$\rho_\infty c_\infty^2$
x,y,z	L
t	L/c_∞

2.3 Boundary Conditions

There are three boundary conditions to be applied to the partial differential equations. The first is a condition of zero mass flux at a solid wall. This takes the form

$$u_i n_i = 0 \;, \tag{2.17}$$

and is applied on any solid surface. The second is that, upstream of the bow-shock, freestream conditions exist. The third is the Kutta condition, applied at the leading-edge of the wing. Numerical implementation of these conditions will be described in Chapter 3.

2.4 Conical Self-Similarity

A flow is conically self-similar if no length scale exists in the radial direction (See Figure 2.2). To look for this class of solutions to the Euler equations, the first step is to change to a conical coordinate system. The (r, η, ζ) system chosen here is related to the Cartesian coordinates by the relations

$$r = \sqrt{x^2 + y^2 + z^2}, \tag{2.18a}$$

$$\eta = \frac{y}{x}, \tag{2.18b}$$

$$\zeta = \frac{z}{x}. \tag{2.18c}$$

Applying this coordinate transformation to the Euler equations (Equation (2.14)) gives

$$\frac{r}{\kappa}\frac{\partial U}{\partial t} + \frac{r}{\kappa^2}\frac{\partial}{\partial r}(F + \eta G + \zeta H) + \frac{\partial}{\partial \eta}(G - \eta F) + \frac{\partial}{\partial \zeta}(H - \zeta F) + 2F = 0 \;, \tag{2.19}$$

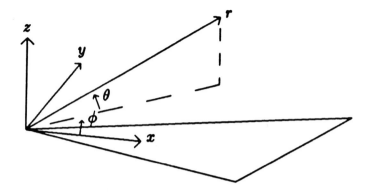

Figure 2.2: Spherical polar coordinate system

where
$$\kappa = \sqrt{1 + \eta^2 + \zeta^2}. \tag{2.20}$$

Expressed in terms of the conical velocities

$$\bar{u} = u + \eta v + \zeta w, \tag{2.21a}$$

$$\bar{v} = v - \eta u, \tag{2.21b}$$

$$\bar{w} = w - \zeta u, \tag{2.21c}$$

the Euler equations become

$$\frac{r}{\kappa}\frac{\partial}{\partial t}\begin{bmatrix}\rho\\\rho u\\\rho v\\\rho w\\\rho E\end{bmatrix} + \frac{r}{\kappa^2}\frac{\partial}{\partial r}\begin{bmatrix}\rho\bar{u}\\\rho u\bar{u}+p\\\rho v\bar{u}+\eta p\\\rho w\bar{u}+\zeta p\\\rho h_0\bar{u}\end{bmatrix} + \frac{\partial}{\partial \eta}\begin{bmatrix}\rho\bar{v}\\\rho u\bar{v}-\eta p\\\rho v\bar{v}+p\\\rho w\bar{v}\\\rho h_0\bar{v}\end{bmatrix} +$$

$$+ \frac{\partial}{\partial \zeta}\begin{bmatrix}\rho\bar{w}\\\rho u\bar{w}-\zeta p\\\rho v\bar{w}\\\rho w\bar{w}+p\\\rho h_0\bar{w}\end{bmatrix} + 2\begin{bmatrix}\rho u\\\rho u^2+p\\\rho uv\\\rho uw\\\rho u h_0\end{bmatrix} = 0. \tag{2.22}$$

Equation (2.22) permits a family of solutions with $\partial/\partial r = 0$ if the boundary conditions satisfy this constraint. This is the conical assumption: that $\partial \mathbf{U}/\partial r = 0$, where $\mathbf{U}$ is the state vector. For the boundary conditions to be conical, they must be applied along rays, i.e. the bow shock, the wing and its leading edges must be generated by rays emanating from the apex. This is only strictly true for supersonic flow with shocks attached at the wing apex. In subsonic flow, or supersonic flow in which the bow shock is not attached at the apex, the far-field boundary condition is not conical. In viscous flows, the deviatoric portion of the stress tensor introduces a $1/x$ factor that prohibits conical self-similarity.

When the conical assumption is applied to Equation (2.22), the second term drops out, leaving

$$\frac{r}{\kappa}\frac{\partial}{\partial t}\begin{bmatrix} \rho \\ \rho u \\ \rho v \\ \rho w \\ \rho E \end{bmatrix} + \frac{\partial}{\partial \eta}\begin{bmatrix} \rho\bar{v} \\ \rho u\bar{v} - \eta p \\ \rho v\bar{v} + p \\ \rho w\bar{v} \\ \rho h_0\bar{v} \end{bmatrix} + \frac{\partial}{\partial \zeta}\begin{bmatrix} \rho\bar{w} \\ \rho u\bar{w} - \zeta p \\ \rho v\bar{w} \\ \rho w\bar{w} + p \\ \rho h_0\bar{w} \end{bmatrix} + 2\begin{bmatrix} \rho u \\ \rho u^2 + p \\ \rho u v \\ \rho u w \\ \rho u h_0 \end{bmatrix} = 0. \qquad (2.23)$$

These are the conical Euler equations. They may be expressed in terms of the state vector $\mathbf{U}$, the Cartesian flux vector $\mathbf{F}$, and conical flux vectors $\hat{\mathbf{G}}$ and $\hat{\mathbf{H}}$ as

$$\frac{r}{\kappa}\frac{\partial \mathbf{U}}{\partial t} + \frac{\partial \hat{\mathbf{G}}}{\partial \eta} + \frac{\partial \hat{\mathbf{H}}}{\partial \zeta} + 2\mathbf{F} = 0. \qquad (2.24)$$

2.5 Topological Considerations

The conical self-similarity assumption has implications on the topology of the flow. Specifically, the entire flow may be characterized by the topology of the flow in a cross-flow plane. The topology of a conical flow is therefore characterized by its conical or cross-flow streamlines. These are the projections of the three-dimensional streamlines onto a spherical surface centered at the apex of the wing. A conical stream-surface, described by $C(\theta, \phi) = 0$, satisfies the relation

$$\vec{u} \cdot \nabla C = 0. \qquad (2.25)$$

This may be written in spherical polar coordinates (See Figure 2.2) as

$$\frac{v}{r}\frac{\partial C}{\partial \theta} + \frac{w}{r\sin\theta}\frac{\partial C}{\partial \phi} = 0, \qquad (2.26)$$

since $\partial C/\partial r = 0$. Thus the angles θ and ϕ along a conical streamline are related by the ordinary differential equation

$$\frac{d\phi}{d\theta} = -\frac{\partial C/\partial \theta}{\partial C/\partial \phi} = \frac{w}{v\sin\theta}. \qquad (2.27)$$

This ODE can have critical points in the region $\theta \in (0, \pi)$, $\phi \in [0, 2\pi)$ only where $d\phi/d\theta$ is indeterminate, i.e. only where $v = w = 0$. If $v\sin\theta$ and w are "well-behaved" near the critical point (θ_0,ϕ_0), i.e. if they can be expanded in Taylor series as

$$v \sin \theta = \frac{\partial (v \sin \theta)}{\partial \theta}(\theta - \theta_0) + \frac{\partial (v \sin \theta)}{\partial \phi}(\phi - \phi_0) + f_1(\theta - \theta_0, \phi - \phi_0), \quad (2.28a)$$

$$w = \frac{\partial w}{\partial \theta}(\theta - \theta_0) + \frac{\partial w}{\partial \phi}(\phi - \phi_0) + f_2(\theta - \theta_0, \phi - \phi_0), \quad (2.28b)$$

where the partial derivatives are evaluated at the critical point and

$$\lim_{\substack{\phi \to \phi_0 \\ \theta \to \theta_0}} \frac{f_1, f_2}{\sqrt{(\phi - \phi_0)^2 + (\theta - \theta_0)^2}} = 0, \quad (2.29)$$

then Equation (2.27) may be linearized about the critical point, giving the parameterized system

$$\frac{d\phi}{dt} = \frac{\partial w}{\partial \phi}(\phi - \phi_0) + \frac{\partial w}{\partial \theta}(\theta - \theta_0), \quad (2.30a)$$

$$\frac{d\theta}{dt} = \frac{\partial (v \sin \theta)}{\partial \phi}(\phi - \phi_0) + \frac{\partial (v \sin \theta)}{\partial \theta}(\theta - \theta_0). \quad (2.30b)$$

This equation set has a solution of the form

$$\phi - \phi_0 = \alpha e^{mt}, \quad (2.31)$$

$$\theta - \theta_0 = \beta e^{mt},$$

where m satisfies

$$m^2 + \left(\frac{\partial w}{\partial \phi} + \frac{\partial (v \sin \theta)}{\partial \theta}\right) m + \left(\frac{\partial w}{\partial \phi}\frac{\partial (v \sin \theta)}{\partial \theta} - \frac{\partial w}{\partial \theta}\frac{\partial (v \sin \theta)}{\partial \phi}\right) = 0. \quad (2.32)$$

The Euler equations may be combined with the continuity equation and the constant total enthalpy condition to give [15]

$$u\left(2 - \frac{v^2 + w^2}{a^2}\right) + v \cot \theta + \frac{\partial v}{\partial \theta}\left(1 - \frac{v^2}{a^2}\right) +$$

$$+ \frac{1}{\sin \theta}\frac{\partial w}{\partial \phi}\left(1 - \frac{w^2}{a^2}\right) - \frac{vw}{a^2}\left(\frac{1}{\sin \theta}\frac{\partial v}{\partial \phi} + \frac{\partial w}{\partial \theta}\right) = 0. \quad (2.33)$$

At the critical point, $v = w = 0$, so this becomes

$$\frac{\partial v}{\partial \theta} + \frac{1}{\sin \theta}\frac{\partial w}{\partial \phi} = -2u. \quad (2.34)$$

Equation (2.32) becomes

$$m^2 - (2u \sin \theta) m + \left(\frac{\partial w}{\partial \phi}\frac{\partial v}{\partial \theta} - \frac{\partial w}{\partial \theta}\frac{\partial v}{\partial \phi}\right) \sin \theta = 0. \quad (2.35)$$

The values for m are then

$$m_1, m_2 = \frac{u \sin \theta}{2} \pm \sqrt{\frac{u^2 \sin^2 \theta}{4} - J} \;, \tag{2.36}$$

where

$$J = \left(\frac{\partial w}{\partial \phi} \frac{\partial v}{\partial \theta} - \frac{\partial w}{\partial \theta} \frac{\partial v}{\partial \phi} \right) \sin \theta \;. \tag{2.37}$$

The different allowable singularities are characterized by the roots m_1 and m_2 [58]. They are:

- Roots are real, distinct and of same sign — the critical point is a node;

- Roots are real, distinct and of opposite sign — the critical point is a saddle point;

- Roots are complex conjugate (but not pure imaginary) — the critical point is a spiral;

- Roots are real and equal — the critical point is a node;

- Roots are pure imaginary — the critical point is a center.

In physical terms, a node on the wing corresponds to a cross-flow stagnation point, with the flow approaching the critical point as it turns downstream. A saddle on the wing corresponds to a line of attachment or separation, with the flow approaching the critical point and then turning away from it. Nodes and saddles may also occur off the wing. The center may be ruled out on physical grounds — it occurs only if $u = 0$, and therefore corresponds to a stagnation line.

It is the spiral singularity that is the most elusive. The spiral point singularity which occurs when the roots are complex conjugate is a logarithmic spiral ($d\phi/d\theta \sim -1/\theta$), and is not seen in conical streamline patterns. Vortex sheet models, such as those of Mangler and Smith [32] and Mangler and Weber [33] predict reciprocal spirals ($d\phi/d\theta \sim -1/\theta^2$).

The solutions presented in this book show nodes, saddles and spirals. The number and locations of the singularities depend upon the flow parameters and wing geometry, and a variety of topologies will be seen.

2.6 Summary

The governing equations for the scheme have been presented. They are the conical Euler equations, restricted to inviscid, supersonic, flow past a conically self-similar body. The non-dimensionalization and boundary conditions for the equations have been described. The implications that the conical assumption has on the flow topology have been determined by examining the first-order singularities of the ordinary differential equation defining a cross-flow stream surface. The physically meaningful singularities are nodes, saddles and spirals.

Chapter 3

Solution Procedure

This chapter outlines the procedure used to obtain numerical solutions of the conical Euler equations, Equation (2.23). The scheme is a finite-volume, multi-stage scheme, in which the state variables are stored at the nodes. It allows for multiple levels of embedded grids. The solution scheme is flow-charted in Figure 3.1. The spatial discretization, added artificial viscosity, temporal discretization and boundary conditions are described in detail below. The data structure used is also discussed.

3.1 Spatial Discretization

This section presents the spatial discretization used to solve the conical Euler equations. The spatial discretization is carried out by generating a grid of points on which the partial differential equations are to be satisfied, and discretizing the equations on this grid. The finite volume formulation and flux integration are similar to those of Hall [20]. The grid generation, finite volume formulation and flux integration are described below.

3.1.1 Grid Generation

The grid generation for embedded grids is carried out in two steps. In the first, a base grid that is logically rectangular is constructed. In the second, the embedded regions are generated. This section begins by describing the first step — the construction of the base grids. Two types of grid generation are used to do this. The first, used to generate the flat plate grids, is an algebraic technique employing a sheared Joukowski transformation. The second, used to generate grids around more general body shapes, is a Poisson equation

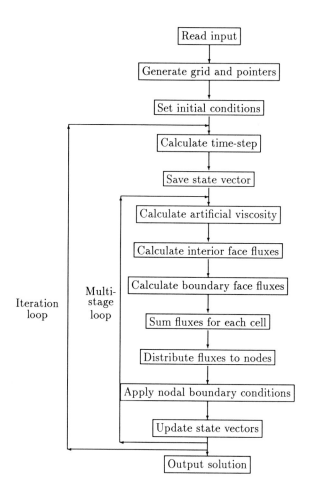

Figure 3.1: Solution scheme flow chart

3.1. SPATIAL DISCRETIZATION

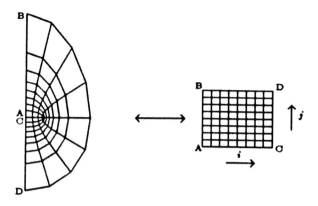

Figure 3.2: Contravariant coordinate system

solver in which grid orthogonality and a given cell aspect ratio are enforced on the inner and outer boundaries of the domain. For both types of grid generation, the computational domain is described as having west-to-east and south-to-north contravariant directions, i and j, so that grid lines are described by $i = constant$ and $j = constant$ (See Figure 3.2).

The grid generation for the flat plate cases is carried out by mapping the (η, ζ) plane to a complex χ plane in which the inner boundary becomes a circle. The mapping used is a Joukowski transformation given by

$$\eta + i\zeta = \chi + \tan^2 \frac{\left(\frac{\pi}{2} - \Lambda\right)}{2\chi} \quad , \tag{3.1}$$

where Λ is the leading-edge sweep of the wing. In the χ plane, $i = constant$ lines are equiangularly spaced rays emanating from the origin and $j = constant$ lines are concentric rings (See Figure 3.3). Grid points are generated along the rays in the χ plane by the formulas

$$r(j) = -\frac{1}{\sigma} \log \left[1 - \left(1 - e^{-\sigma}\right) \frac{j - j^{(in)}}{j^{(out)} - j^{(in)}} \right] \quad , \tag{3.2a}$$

$$\chi(j) = \chi(j^{(in)}) + \left[\chi(j^{(out)}) - \chi(j^{(in)})\right] r(j) \quad , \tag{3.2b}$$

where σ is a stretching parameter, and $j^{(in)}$ and $j^{(out)}$ are the values of j on the inner and outer boundaries of the domain. This method, originally proposed by Marconi [34], yields

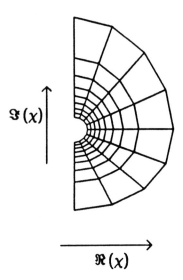

Figure 3.3: Grid in transformed plane

near-conformal grids with good resolution near the wing. An example of a grid generated by this method is shown in Figure 3.4.

The grid generation for more general bodies is carried out by a method similar to one originally proposed by Steger and Sorenson [60]. The method consists of solving simultaneous Poisson equations

$$\frac{\partial^2 i}{\partial \eta^2} + \frac{\partial^2 i}{\partial \zeta^2} = P(\eta, \zeta) \;, \tag{3.3a}$$

$$\frac{\partial^2 j}{\partial \eta^2} + \frac{\partial^2 j}{\partial \zeta^2} = Q(\eta, \zeta) \;, \tag{3.3b}$$

where P and Q are source terms calculated by a method to be described.

The equations are transformed to the computational space by interchanging the dependent and independent variables, giving [2]

$$\alpha \frac{\partial^2 \eta}{\partial i^2} - 2\beta \frac{\partial^2 \eta}{\partial i \partial j} + \gamma \frac{\partial^2 \eta}{\partial j^2} = -J^2 \left[P \frac{\partial \eta}{\partial i} + Q \frac{\partial \eta}{\partial j} \right] \;, \tag{3.4a}$$

$$\alpha \frac{\partial^2 \zeta}{\partial i^2} - 2\beta \frac{\partial^2 \zeta}{\partial i \partial j} + \gamma \frac{\partial^2 \zeta}{\partial j^2} = -J^2 \left[P \frac{\partial \zeta}{\partial i} + Q \frac{\partial \zeta}{\partial j} \right] \;, \tag{3.4b}$$

where

$$\alpha = \left(\frac{\partial \eta}{\partial j}\right)^2 + \left(\frac{\partial \zeta}{\partial j}\right)^2 \;, \tag{3.5a}$$

3.1. SPATIAL DISCRETIZATION

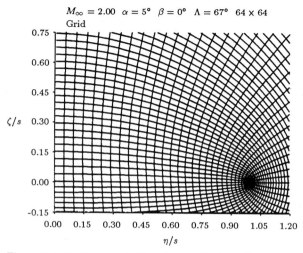

Figure 3.4: Grid generated by sheared Joukowski transformation

$$\beta = \frac{\partial \eta}{\partial i}\frac{\partial \eta}{\partial j} + \frac{\partial \zeta}{\partial i}\frac{\partial \zeta}{\partial j}, \quad (3.5b)$$

$$\gamma = \left(\frac{\partial \eta}{\partial i}\right)^2 + \left(\frac{\partial \zeta}{\partial i}\right)^2, \quad (3.5c)$$

$$J = \frac{\partial \zeta}{\partial i}\frac{\partial \eta}{\partial j} - \frac{\partial \zeta}{\partial j}\frac{\partial \eta}{\partial i}. \quad (3.5d)$$

Now the boundary conditions are to be specified in terms of $\eta(i,j)$ and $\zeta(i,j)$ on the boundaries (Dirichlet conditions on $j = constant$ boundaries, Neumann conditions on $i = constant$ boundaries). The equations are solved by a successive line over-relaxation (SLOR) iterative procedure.

The source terms P and Q are calculated so that:

1. Grid lines are orthogonal to the inner and outer boundaries;

2. Boundary cells have specified aspect ratios $\mathcal{R}^{(in)}$ and $\mathcal{R}^{(out)}$.

Boundary source terms $P^{(in)}$, $Q^{(in)}$, $P^{(out)}$ and $Q^{(out)}$ that enforce these constraints are calculated, and the source terms in the interior of the field are composed of a combination of the boundary terms, that is

$$P(i,j) = P^{(in)}(i)e^{-a|j-j^{(in)}|} + P^{(out)}(i)e^{-a|j-j^{(out)}|}, \quad (3.6a)$$

$$Q(i,j) = Q^{(in)}(i)e^{-a|j-j^{(in)}|} + Q^{(out)}(i)e^{-a|j-j^{(out)}|}, \quad (3.6b)$$

where a is a positive constant of order one.

The boundary source terms are calculated by evaluating the transformed Poisson equations, Equations (3.4a) and (3.4b), on the boundaries and solving them for P and Q there. Since the boundary point distributions $\eta(i, j^{(in)})$, $\zeta(i, j^{(in)})$, $\eta(i, j^{(out)})$ and $\zeta(i, j^{(out)})$ are known, the first and second derivatives in i of these distributions are also known. This leaves the j derivatives to be determined. They are set by the orthogonality and cell aspect ratio constraints. The orthogonality constraint is

$$\nabla i \cdot \nabla j = 0 \tag{3.7}$$

or, exchanging the dependent and independent variables,

$$\frac{\partial \eta}{\partial i}\frac{\partial \eta}{\partial j} + \frac{\partial \zeta}{\partial i}\frac{\partial \zeta}{\partial j} = 0. \tag{3.8}$$

The cell aspect ratio constraint is

$$\mathcal{R} = \frac{\partial s/\partial j}{\partial s/\partial i}, \tag{3.9}$$

so that

$$\mathcal{R}^{(in)} = \left[\frac{\left(\frac{\partial \eta}{\partial j}\right)^2 + \left(\frac{\partial \zeta}{\partial j}\right)^2}{\left(\frac{\partial \eta}{\partial i}\right)^2 + \left(\frac{\partial \zeta}{\partial i}\right)^2}\right]^{\frac{1}{2}}\Bigg|_{j^{(in)}}, \tag{3.10a}$$

$$\mathcal{R}^{(out)} = \left[\frac{\left(\frac{\partial \eta}{\partial j}\right)^2 + \left(\frac{\partial \zeta}{\partial j}\right)^2}{\left(\frac{\partial \eta}{\partial i}\right)^2 + \left(\frac{\partial \zeta}{\partial i}\right)^2}\right]^{\frac{1}{2}}\Bigg|_{j^{(out)}}. \tag{3.10b}$$

Combining Equations (3.8), (3.10a) and (3.10b), the first derivatives in j may be determined. They are

$$\frac{\partial \eta}{\partial j}(i, j^{(in)}) = -\frac{\partial \zeta}{\partial i}(i, j^{(in)})\, \mathcal{R}^{(in)}, \tag{3.11a}$$

$$\frac{\partial \zeta}{\partial j}(i, j^{(in)}) = +\frac{\partial \eta}{\partial i}(i, j^{(in)})\, \mathcal{R}^{(in)}, \tag{3.11b}$$

$$\frac{\partial \eta}{\partial j}(i, j^{(out)}) = -\frac{\partial \zeta}{\partial i}(i, j^{(out)})\, \mathcal{R}^{(out)}, \tag{3.11c}$$

$$\frac{\partial \zeta}{\partial j}(i, j^{(out)}) = +\frac{\partial \eta}{\partial i}(i, j^{(out)})\, \mathcal{R}^{(out)}. \tag{3.11d}$$

The second derivative in j is obtained by a one-sided difference on the previous iteration for the grid.

3.1. SPATIAL DISCRETIZATION

The inner and outer boundary source terms are therefore given by

$$P^{(in)}(i) = \frac{1}{J}\left(\frac{\partial \zeta}{\partial j}(i,j^{(in)})R_\eta^{(in)}(i) - \frac{\partial \eta}{\partial j}(i,j^{(in)})R_\zeta^{(in)}(i)\right), \quad (3.12a)$$

$$Q^{(in)}(i) = \frac{1}{J}\left(-\frac{\partial \zeta}{\partial i}(i,j^{(in)})R_\eta^{(in)}(i) + \frac{\partial \eta}{\partial i}(i,j^{(in)})R_\zeta^{(in)}(i)\right), \quad (3.12b)$$

$$P^{(out)}(i) = \frac{1}{J}\left(\frac{\partial \zeta}{\partial j}(i,j^{(out)})R_\eta^{(out)}(i) - \frac{\partial \eta}{\partial j}(i,j^{(out)})R_\zeta^{(out)}(i)\right), \quad (3.12c)$$

$$Q^{(out)}(i) = \frac{1}{J}\left(-\frac{\partial \zeta}{\partial i}(i,j^{(out)})R_\eta^{(out)}(i) + \frac{\partial \eta}{\partial i}(i,j^{(out)})R_\zeta^{(out)}(i)\right), \quad (3.12d)$$

where

$$R_\eta^{(in)}(i) = \frac{1}{J^2}\left[\alpha\frac{\partial^2 \eta}{\partial i^2} - 2\beta\frac{\partial^2 \eta}{\partial i \partial j} + \gamma\frac{\partial^2 \eta}{\partial j^2}\right]\bigg|_{(i,j^{(in)})}, \quad (3.13a)$$

$$R_\zeta^{(in)}(i) = \frac{1}{J^2}\left[\alpha\frac{\partial^2 \zeta}{\partial i^2} - 2\beta\frac{\partial^2 \zeta}{\partial i \partial j} + \gamma\frac{\partial^2 \zeta}{\partial j^2}\right]\bigg|_{(i,j^{(in)})}, \quad (3.13b)$$

$$R_\eta^{(out)}(i) = \frac{1}{J^2}\left[\alpha\frac{\partial^2 \eta}{\partial i^2} - 2\beta\frac{\partial^2 \eta}{\partial i \partial j} + \gamma\frac{\partial^2 \eta}{\partial j^2}\right]\bigg|_{(i,j^{(out)})}, \quad (3.13c)$$

$$R_\zeta^{(out)}(i) = \frac{1}{J^2}\left[\alpha\frac{\partial^2 \zeta}{\partial i^2} - 2\beta\frac{\partial^2 \zeta}{\partial i \partial j} + \gamma\frac{\partial^2 \zeta}{\partial j^2}\right]\bigg|_{(i,j^{(out)})}, \quad (3.13d)$$

and the j derivatives of η and ζ are calculated as described above. The transformed Poisson equations are solved in delta form by a SLOR procedure in which the source terms P and Q are under-relaxed (typically by a factor of order 10^{-2}). The initial guess for the grid is constructed of equally-spaced points along rays from the inner to outer boundaries. A typical grid obtained by this procedure is shown in Figure 3.5.

The generation of embedding regions is carried out by the procedure described below. Starting with the base grid, a rectangular region is chosen for the first level of embedding. Each cell within this region is divided into quarters. Nodes are inserted at the midpoint of each face of the cell to be divided (on boundary faces the new node position is calculated from the analytic expression defining the boundary), and a node is added at the geometric center of the cell to be divided. The next level of embedding is carried out by choosing a rectangle within the first embedded region, and dividing all these cells into quarters. This process is repeated until the desired level of embedding is reached. Obviously the nodes may not be numbered in an (i,j) manner on a grid of this type; the data structure chosen for the scheme is described in Section 3.6.

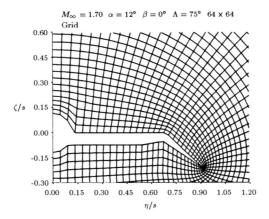

Figure 3.5: Grid generated by Poisson method

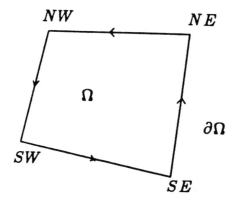

Figure 3.6: Finite volume integration

3.1.2 Finite Volume Formulation

This section describes the discretization of the partial differential equations on the grid. The discretization is formulated by integrating the conical Euler equations, Equation (2.23), over a cell (See Figure 3.6). This gives

$$\iint_\Omega \frac{r}{\kappa}\frac{\partial \mathbf{U}}{\partial t} d^2 x + \iint_\Omega \left[\frac{\partial \hat{\mathbf{G}}}{\partial \eta} + \frac{\partial \hat{\mathbf{H}}}{\partial \zeta}\right] d^2 x + \iint_\Omega 2\mathbf{F} d^2 x = 0. \quad (3.14)$$

Using Gauss' theorem and the mean value theorem, this may be rewritten as

$$A \overline{\frac{r}{\kappa}\frac{\partial \mathbf{U}}{\partial t}} + \oint_{\partial \Omega} \left[\hat{\mathbf{G}} d\zeta - \hat{\mathbf{H}} d\eta\right] + 2A\overline{\mathbf{F}} = 0 \quad , \quad (3.15)$$

3.2. ARTIFICIAL VISCOSITY

where an overbar denotes an average over the cell. Solving for $\overline{\partial \mathbf{U}/\partial t}$, the time rate of change of the state vector within a cell, gives

$$\frac{\overline{\partial \mathbf{U}}}{\partial t} = -\frac{\overline{\kappa}}{rA} \left[\oint_{\partial \Omega} \left[\hat{\mathbf{G}} d\zeta - \hat{\mathbf{H}} d\eta \right] + 2A\overline{\mathbf{F}} \right]. \tag{3.16}$$

3.1.3 Flux Integration

This section deals with the calculation of the right-hand side of Equation (3.16). There are two terms to be calculated: the line integral of the fluxes, and the cell-average of the source term. The line integral is calculated by a trapezoidal integration around the cell, i.e.

$$\oint_{\partial \Omega} \left[\hat{\mathbf{G}} d\zeta - \hat{\mathbf{H}} d\eta \right] = \sum_{faces} \left[\frac{1}{2}(\hat{\mathbf{G}}_1 + \hat{\mathbf{G}}_2)(\zeta_2 - \zeta_1) - \frac{1}{2}(\hat{\mathbf{H}}_1 + \hat{\mathbf{H}}_2)(\eta_2 - \eta_1) \right], \tag{3.17}$$

where the subscripts denote the two nodes that define the face, ordered so that the integral is carried out in a counter-clockwise sense. This is formally second-order accurate, regardless of the stretching of the grid [64].

The cell-average of the source term is calculated by

$$\overline{\mathbf{F}} = \frac{1}{4}(\mathbf{F}_{sw} + \mathbf{F}_{se} + \mathbf{F}_{ne} + \mathbf{F}_{nw}), \tag{3.18}$$

where the subscripts denote the nodes defining the cell (See Figure 3.6). The cell area A is calculated by the formula

$$A = \frac{1}{2} \left[(\zeta_{ne} - \zeta_{sw})(\eta_{se} - \eta_{nw}) - (\eta_{ne} - \eta_{sw})(\zeta_{se} - \zeta_{nw}) \right]. \tag{3.19}$$

The remaining factor in the residual equation, $\overline{\kappa/rA}$, will be included in the time step calculation, discussed in Section 3.3.

3.2 Artificial Viscosity

The flux integration described above is equivalent to a centered difference approximation for the derivatives of the flux vectors. The centered difference approximation to $\partial u/\partial x$ is given by

$$\frac{u(x + \Delta x) - u(x - \Delta x)}{2\Delta x} = \frac{1}{2\Delta x} \left[u + \frac{\partial u}{\partial x} \Delta x + \frac{\Delta x^2}{2} \frac{\partial^2 u}{\partial x^2} + \frac{\Delta x^3}{6} \frac{\partial^3 u}{\partial x^3} + \ldots \right] -$$

$$-\frac{1}{2\Delta x}\left[u - \frac{\partial u}{\partial x}\Delta x + \frac{\Delta x^2}{2}\frac{\partial^2 u}{\partial x^2} - \frac{\Delta x^3}{6}\frac{\partial^3 u}{\partial x^3} + \ldots\right] =$$
$$= \frac{1}{2\Delta x}\left[2\frac{\partial u}{\partial x}\Delta x + \frac{\Delta x^3}{3}\frac{\partial^3 u}{\partial x^3} + \ldots\right] =$$
$$= \frac{\partial u}{\partial x} + \frac{\Delta x^2}{6}\frac{\partial^3 u}{\partial x^3} + \ldots, \quad (3.20)$$

which is dispersive, due to the odd-order derivative in the leading-order term of the truncation error. To damp high frequency error modes, higher-order diffusive terms are added to the difference equations. This is referred to as artificial viscosity.

The artificial viscosity is based on one proposed by Rizzi and Eriksson [55] and one proposed by Ni [46]. It is composed of two elements: a nonlinear second-difference and a linear fourth-difference. The second-difference term is designed to mimic physical viscosity, allowing the Euler equations to "capture" discontinuities. The fourth-difference is used to counteract aliasing errors and dissipate the high frequency error modes that occur even in low-gradient regions of the flow. The second-difference term is constructed so as to be first-order in high-gradient regions and small elsewhere; the fourth-difference term is third-order everywhere.

The fourth-difference operator is given by

$$D_4(\hat{\mathbf{U}}) = \epsilon_4 L^2(\hat{\mathbf{U}}). \quad (3.21)$$

In the above, $\hat{\mathbf{U}}$ is a modified state vector in which the energy term ρE is replaced by ρh_0 so that the discrete equations permit a solution with constant total enthalpy. The fourth-difference damping coefficient, ϵ_4, is of order 10^{-3}. Justification of this choice of order of magnitude is given in the stability analysis in Appendix A. L is an unweighted Laplacian operator and L^2 a biharmonic operator. In the interior, away from boundaries and embedding interfaces, the Laplacian takes the form (See Figure 3.7)

$$L(\hat{\mathbf{U}}) = \hat{\mathbf{U}}_{SW} + \hat{\mathbf{U}}_{SE} + \hat{\mathbf{U}}_{NE} + \hat{\mathbf{U}}_{NW} +$$
$$+ 2\hat{\mathbf{U}}_S + 2\hat{\mathbf{U}}_E + 2\hat{\mathbf{U}}_N + 2\hat{\mathbf{U}}_W - 12\hat{\mathbf{U}}. \quad (3.22)$$

The way that the Laplacian is calculated and the form that it takes on boundaries and at interfaces will be described in Section 3.5.

The second-difference operator is nonlinear. It is of the form

$$D_2(\hat{\mathbf{U}}) = \epsilon_2 \overline{L}(\hat{\mathbf{U}}, \delta p). \quad (3.23)$$

3.2. ARTIFICIAL VISCOSITY

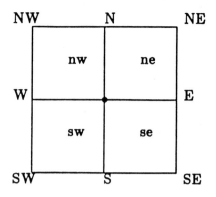

Figure 3.7: Second-difference damping stencil

Again, the damping operator acts on the modified state vector, $\hat{U}$. The second-difference damping coefficient, ϵ_2, is of order 10^{-2}. Justification of this order of magnitude is given in Appendix A. $\overline{L}$ is a modified Laplacian operator, using a weighting function δp in the following manner (See Figure 3.7)

$$\begin{aligned}\overline{L}(\hat{U}, \delta p) &= (\hat{U}_{SW} - \hat{U})\delta p_{sw} + (\hat{U}_{SE} - \hat{U})\delta p_{se} + (\hat{U}_{NE} - \hat{U})\delta p_{ne} \; + \\ &+ (\hat{U}_{NW} - \hat{U})\delta p_{nw} + 2(\hat{U}_S - \hat{U})\delta p_s + 2(\hat{U}_E - \hat{U})\delta p_e \; + \\ &+ 2(\hat{U}_N - \hat{U})\delta p_n + 2(\hat{U}_W - \hat{U})\delta p_w. \end{aligned} \quad (3.24)$$

This is the form of $\overline{L}$ in the interior. The way in which it is calculated, and the form that it takes at boundaries and interfaces, will be described in Section 3.5. The form of the weighting function δp is chosen to give good shock and vortex resolution. The value of δp at a node is given by is given by

$$\delta p = \left| \frac{L(p)/p}{||L(p)/p||_\infty} \right|, \quad (3.25)$$

where $|| \; ||_\infty$ denotes the L_∞ or max norm, so that $0 < \delta p < 1$. The values for δp between nodes are obtained by taking the maximum, for example

$$\delta p_{sw} = \max\left(\delta p, \delta p_{SW}\right). \quad (3.26)$$

Values for δp at boundaries and interfaces are obtained by a first-order extrapolation from the interior.

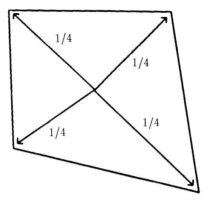

Figure 3.8: Distribution scheme

3.3 Temporal Discretization

Equation (3.16) is a semi-discrete equation for $\overline{\partial \mathbf{U}/\partial t}$, the time rate of change of the state vector within a cell. This forms a system of coupled nonlinear ordinary differential equations. Two processes are needed to integrate them: one to distribute $\overline{\partial \mathbf{U}/\partial t}$ to the nodes, and one to integrate the changes at the nodes. For the first, a simple $\left(\frac{1}{4}, \frac{1}{4}, \frac{1}{4}, \frac{1}{4}\right)$ distribution is used (See Figure 3.8). For the second, a multi-stage scheme is used.

The multi-stage scheme is given by, for iteration n [24],

$$\mathbf{U}^{(0)} = \mathbf{U}^n , \tag{3.27a}$$

$$\mathbf{U}^{(1)} = \mathbf{U}^{(0)} - \alpha_1 \left[\Delta t \overline{\frac{\kappa}{rA}} \mathbf{C}^{(0)} - \lambda \mathbf{D}^{(0)} \right] , \tag{3.27b}$$

$$\mathbf{U}^{(2)} = \mathbf{U}^{(0)} - \alpha_2 \left[\Delta t \overline{\frac{\kappa}{rA}} \mathbf{C}^{(1)} - \lambda \mathbf{D}^{(1)} \right] , \tag{3.27c}$$

$$\vdots$$

$$\mathbf{U}^{(k)} = \mathbf{U}^{(0)} - \alpha_k \left[\Delta t \overline{\frac{\kappa}{rA}} \mathbf{C}^{(k-1)} - \lambda \mathbf{D}^{(k-1)} \right] , \tag{3.27d}$$

$$\mathbf{U}^{n+1} = \mathbf{U}^{(k)} . \tag{3.27e}$$

In the above, $\mathbf{C}$ is the convective operator, given by

$$\mathbf{C}^{(k)} = \oint_{\partial \Omega} \left[\hat{\mathbf{G}}(\mathbf{U}^{(k)}) d\zeta - \hat{\mathbf{H}}(\mathbf{U}^{(k)}) d\eta \right] + 2A \bar{\mathbf{F}}(\mathbf{U}^{(k)}) , \tag{3.28}$$

and $\mathbf{D}$ is the damping operator, given by

$$\mathbf{D}^{(k)} = \mathbf{D}_2(\hat{\mathbf{U}}^{(k)}) - \mathbf{D}_4(\hat{\mathbf{U}}^{(k)}). \tag{3.29}$$

3.4. BOUNDARY CONDITIONS

The modified time-step factor, $\Delta t \,\overline{\kappa/rA}$, is given by

$$\Delta t \overline{\frac{\kappa}{rA}} = \lambda \overline{\frac{\kappa}{r}} \max_{faces} \left[\frac{1}{u_i n_i + c\sqrt{n_i n_i}} \right]. \tag{3.30}$$

where λ is the CFL number, u_i is the ith component of the velocity at the face, c is the speed of sound at the face and n_i is the ith component of the non-normalized face normal. Derivation of this time step is given in Appendix A. The radius r is arbitrarily set to unity for the time-step calculation. For the cases in this book, a four-stage scheme was chosen, with multi-stage coefficients of

$$\alpha_1 = .25\,, \quad \alpha_2 = .33\,, \quad \alpha_3 = .50\,, \quad \alpha_4 = 1.0\,. \tag{3.31}$$

3.4 Boundary Conditions

As stated in Chapter 2, there are three boundary conditions to be applied to the Euler equations. They are:

1. No flux at the solid wall;

2. Free-stream conditions upstream of the bow shock;

3. The Kutta condition at the leading-edges.

The numerical scheme introduces non-physical boundaries — the symmetry plane and the embedding interfaces — which must be treated as well. The numerical implementation of the boundary conditions at the three physical boundaries and the two artificial boundaries will be described below.

3.4.1 Solid Wall Condition

The physical boundary condition at the wall is that

$$u_i n_i = 0. \tag{3.32}$$

This has a direct analog in the computational plane. It is achieved by altering the flux integration for the cell faces on the wall, given by

$$\text{Flux} \;=\; \int \left[\hat{\mathbf{G}} d\zeta - \hat{\mathbf{H}} d\eta \right] \;=\; \tag{3.33a}$$

$$= \begin{bmatrix} \int \rho u_n \\ \int [\rho u u_n - p(\eta d\zeta - \zeta d\eta)] \\ \int [\rho v u_n + p d\zeta] \\ \int [\rho w u_n - p d\eta] \\ \int \rho h_0 u_n \end{bmatrix}, \qquad (3.33b)$$

where u_n is the velocity normal to the wall, given by

$$u_n = \overline{v} d\zeta - \overline{w} d\eta. \qquad (3.34)$$

The no-flux condition is enforced by zeroing the u_n terms, giving

$$\mathbf{Flux} = \begin{bmatrix} 0 \\ \int [-p(\eta d\zeta - \zeta d\eta)] \\ \int p d\zeta \\ \int -p d\eta \\ 0 \end{bmatrix}, \qquad (3.35)$$

where the integral is evaluated by a trapezoidal integration, as before.

In addition to this change in the flux integration at the solid wall, the flux distribution is changed to account for the fact that each wall node belongs to half as many cells as it would if it were not on the wall. The distributed flux to each wall node is therefore doubled.

3.4.2 Far-Field Condition

The far-field condition is that free-stream conditions exist upstream of the bow shock. This is implemented numerically by placing the outer boundary of the computational plane outside the bow shock and setting the state vector at each point on the outer boundary to the free-stream state vector. The outer boundary chosen is an expanded Mach cone, shifted to leeward to account for the angle of attack. That is,

$$\eta = r_s \sin\varphi \;, \qquad (3.36a)$$

$$\zeta = r_s \cos\varphi + z_s \cos^2\varphi \;, \qquad (3.36b)$$

where

$$r_s = 3\sin^{-1}\left(\frac{1}{M_\infty}\right) \;, \qquad (3.37a)$$

$$z_s = \max(3\sin\alpha, 3\sin 10°). \qquad (3.37b)$$

3.4. BOUNDARY CONDITIONS

3.4.3 Kutta Condition

Physically the Kutta condition implies that the cross-flow leaves the leading-edge smoothly. This rules out the possibility of infinite cross-flow velocity at the leading-edge. In a real flow, it is the viscosity that sets the Kutta condition. Similarly, it is found that in computations with the present scheme, artificial viscosity in the region of the leading-edge enforces the Kutta condition. Thus the Kutta condition is enforced by arbitrarily setting the weighting function δp for the second-difference damping to a value of one for several (less that ten) nodes in the vicinity of the leading-edge.

3.4.4 Symmetry Plane Condition

The symmetry plane is introduced so that the flow past a wing at zero yaw may be solved on a half-plane. The condition to be enforced at the symmetry line is that the through-flow velocity, v, is zero. This is implemented by setting v to zero initially and zeroing the distributed flux for the y-direction momentum equation at each iteration. As with the wall nodes, the remaining distributed fluxes are doubled.

3.4.5 Embedding Interfaces

Interfaces between two levels of grid refinement lead to cells of more than four nodes and faces. The embedding philosophy used in this work is a very simple one. It restricts the grid to a series of nested rectangles (See Figure 3.9) in which there is refinement by a factor of two in each direction at the interfaces. This is more restrictive than the grids used by Dannenhoffer and Baron [12], Kallinderis and Baron [26], and Shapiro and Murman [57]. With this restriction, only one type of interface cell is generated — a cell with five faces and five nodes. These cells are the coarse cells which abut the embedding interface. This makes the interface treatment very simple. The flux integration for these cells is carried out exactly as above, with five faces for the cell instead of the usual four. The distribution step is modified, however, so that the "extra" node is updated. The distribution used is diagrammed in Figure 3.10.

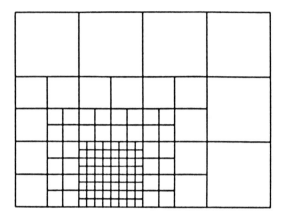

Figure 3.9: Embedding philosophy

$$a = \frac{1}{4+2\sqrt{2}}$$

$$b = \frac{\sqrt{2}}{4+2\sqrt{2}}$$

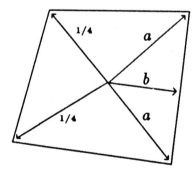

Figure 3.10: Distribution scheme at interfaces

3.5 Artificial Viscosity Boundary Conditions

Each of the physical and numerical boundaries described above requires a special treatment of the artificial viscosity. The underlying philosophy taken here is to preserve conservation. This makes the aritificial viscosity locally dispersive rather than dissipative at boundaries and interfaces; however, it remains globally dissipative [14].

The stencil for the unweighted Laplacian operator, L, away from boundaries, is shown in Figure 3.11. To avoid special treatment at boundaries, the Laplacian may be calculated on a cell-to-cell basis. The value for $L(\hat{\mathbf{U}}_i)$ at a node i is expressed as contributions from the n cells which contain that node,

$$L(\hat{\mathbf{U}}_i) = -4n\hat{\mathbf{U}}_i + \sum_{cell\,1}^{n} \sum_{node\,1}^{4} \hat{\mathbf{U}}. \tag{3.38}$$

If the Laplacian is calculated in this way, conservation is preserved, even at boundaries and interfaces. At a node on the solid wall, far-field or symmetry boundary, the stencil takes the form shown in Figure 3.12. At a corner node, the stencil takes the form shown in Figure 3.13. The embedding interfaces introduce three types of nodes where the Laplacian stencil is altered: corner nodes (interface nodes which belong to one fine cell and three coarse cells), "hanging" nodes (interface nodes which belong to two fine cells and one coarse cell), and typical interface nodes (which belong to two coarse cells and two fine cells). The L stencils for these nodes are shown in Figures 3.14-3.16 .

The fourth-difference damping is based on the operator L^2. It is obtained by applying the Laplacian operator twice, in the conservative form given above. Again, conservation is enforced, but the stencils are altered at the boundaries and interfaces and at nodes that are neighbors of boundary or interface nodes. The altered stencils are not strictly dissipative, but are conservative.

The weighted Laplacian operator, $\bar{L}$, is calculated in a different manner, but is also conservative. The value for $\bar{L}_i$ at a node i is made up of contributions from each face and cell that contain the node. It is given by

$$\bar{L}_i = \sum_{faces} \max\left(\delta p_i, \delta p_j\right) [\mathbf{U}_j - \mathbf{U}_i] + \sum_{cells} \max\left(\delta p_i, \delta p_k\right) [\mathbf{U}_k - \mathbf{U}_i], \tag{3.39}$$

where each face is defined by nodes i and j, and nodes i and k are at opposite corners of the cell (See Figure 3.17). As with the unweighted Laplacian, this procedure gives a dispersive stencil at boundaries and embedding interfaces, but is everywhere conservative.

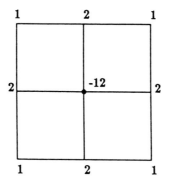

Figure 3.11: Laplace stencil at interior node

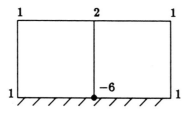

Figure 3.12: Laplace stencil at boundary node

3.5. ARTIFICIAL VISCOSITY BOUNDARY CONDITIONS

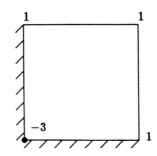

Figure 3.13: Laplace stencil at corner node

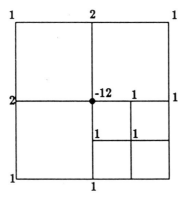

Figure 3.14: Laplace stencil at interface corner node

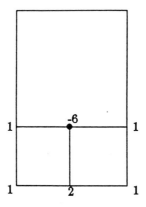

Figure 3.15: Laplace stencil at interface "hanging" node

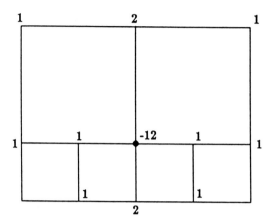

Figure 3.16: Laplace stencil at typical interface node

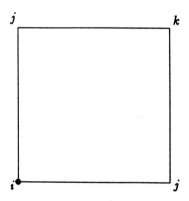

Figure 3.17: Weighted Laplacian calculation

3.6 Data Structure

Because of the embedded regions of the grid, the grid points may not be indexed in a typical (i,j) fashion [1,12,65]. Instead, nodes, cells and faces are singly indexed, and pointer vectors are constructed with the necessary information relating nodes, cells and faces. The pointer vectors used in this scheme are:

1. A cell-to-node pointer, relating a cell to the four nodes that define it;

2. A cell-to-face pointer, relating a cell to its four faces;

3. A face-to-node pointer, relating a face to the two nodes defining it;

4. A node-to-node pointer, relating a node to its (at most) four neighbors.

Each pointer has a specific use. The cell-to-node pointer is used in the flux distribution, the damping calculation and the source term averaging. In the distribution step, for instance, each cell is visited and the flux in that cell is distributed to the nodes of the cell. The source term averaging is carried out by visiting a cell and averaging the values of the source term at the four nodes defining the cell. The damping calculation is similar to the distribution step, in that cells are visited, and changes in the cell are distributed to the nodes of the cell. The cell-to-node pointer is thus necessary for each of these procedures, as it defines the nodes which make up the cell.

The cell-to-face pointer is used in the flux summation step. Each cell is visited, and the fluxes from the four faces defining the cell are summed. The face-to-node pointer is used in the calculation of the face fluxes. Each face is visited, and the coordinates and the values of the flux vectors at the nodes defining the face are used to carry out the trapezoidal integration. The node-to-node pointer is used to calculate grid metrics and cell areas, and to extrapolate quantities (such as the time step and the weighting function δp) to the boundaries and interfaces.

In addition to these pointers, pointers which contain boundary information are necessary. They are:

1. A vector containing the index of each solid wall face;

2. A vector containing the index of each solid wall node;

3. A vector containing the index of each periodic boundary node;

4. A vector containing the index of each far-field node;

5. A vector containing the index of each node at which the weighting factor for the damping is to be overridden;

6. A vector containing the index of each coarse cell which abuts an embedding interface;

7. A vector relating each of these interface cells to its "extra" node.

8. A vector relating each of these interface cells to its "extra" face;

9. A vector containing the orientation of each of these "extra" faces (i.e. which side of the cell the face is on);

The boundary conditions at physical and numerical boundaries are enforced using these pointers. The wall boundary condition is enforced by visiting each face on the wall, and altering the flux there. The far-field boundary condition is enforced by visiting each far-field node and setting the state vector there to the free-stream state vector. The Kutta condition is enforced by overriding the weighting factor δp at the nodes near the leading-edge. The interfaces are handled by visiting each coarse cell that abuts an embedding interface, adding the flux from the "extra" face there, and distributing a portion of the flux to the "extra" node.

3.7 Summary

This chapter has presented the solution algorithm used to solve the conical Euler equations. Two types of grid generation for the base grid have been described: a sheared Joukowski transformation used for flat plate grids, and a Poisson equation method with specially constructed source terms for more general grids. The generation of the embedded regions has also been discussed. The governing equations have been discretized by a finite volume formulation. A trapezoidal integration has been used for the cross-flow flux line integral evaluation, and a simple averaging for the conical source term area integral evaluation. The artificial viscosity is a blend of a nonlinear second-difference and a linear fourth-difference, designed to be conservative at boundaries and interfaces. The temporal discretization consists of a simple distribution to the nodes, modified at interfaces, and

3.7. SUMMARY

a multi-stage integration in time. A simple and efficient data structure to relate nodes, cells and faces and to contain boundary and interface notation has been presented.

Chapter 4

Basic Characteristics of Solutions

This chapter describes the output from the solution scheme presented in Chapter 3. The different flow variables to be examined will be presented, and the way that they represent the flow features will be explained. Representative plots of each of the variables will be presented and described. Two cases will be used to illustrate the output: a low Mach number and angle of attack case ($M_\infty = 1.1$, $\alpha = 10°$, $\Lambda = 75°$) and a higher Mach number and angle of attack case ($M_\infty = 1.95$, $\alpha = 25°$, $\Lambda = 75°$).

All of the plots in this book are cross-flow plots. They are the projections of the $r = 1$ spherical shell onto a plane. The coordinate axes are labelled η/s and ζ/s, where s is the local semi-span of the delta wing and $\eta = y/x$ and $\zeta = z/x$ are the conical cross-flow variables. The wing is at $\zeta/s = 0$ and extends from $\eta/s = -1$ to $\eta/s = 1$. For the symmetric cases, only the positive η plane is shown. For the vortex flap cases, s is taken to be the local semi-span with the flap undeflected. Each plot has a title which contains the free-stream Mach number (M_∞), the angle of attack (α), the angle of yaw (β), the leading-edge sweep (Λ) and the equivalent global grid refinement for the case. Contour plots also contain the contour increment in the upper left-hand corner of the plot.

4.1 Topological Aspects of the Solutions

Topological information about the flow is presented through cross-flow velocity vector plots, cross-flow streamline plots, flow angularity plots and tuft plots. Both the velocity vector and streamline plots are based on the cross-flow velocities $\bar{v} = v - \eta u$ and $\bar{w} = w - \zeta u$. The tuft and angularity plots are based on the Cartesian velocity components.

44   CHAPTER 4. BASIC CHARACTERISTICS OF SOLUTIONS

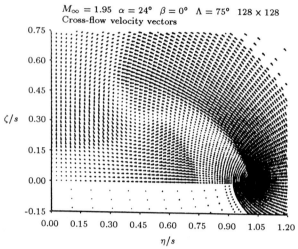

Figure 4.1: Cross-flow velocity vectors

The cross-flow velocity vector plots are generated by drawing, at each node, an arrow whose vector components are proportional to the local values of $\bar{v}$ and $\bar{w}$. Figures 4.1-4.4 show the cross-flow velocity vectors for the case with high free-stream Mach number and angle of attack. Figure 4.1 shows the entire vortex region. The feeding sheet and vortex may be seen, as well as three cross-flow shocks: a shock above the vortex, where the flow direction is inboard; a shock inboard of the vortex, where the flow direction is towards the wing; and a shock under the vortex, where the flow is outboard. This reverse cross-flow shock under the vortex, the existence of which was first postulated by Vorropoulos and Wendt [67] appears in virtually every flow calculated. The shocks above and inboard of the vortex are representative of high angle of attack cases. Figures 4.2-4.4 show blowups of the vortex core, the feeding sheet and the reverse cross-flow shock, respectively. The core (Figure 4.2) is characterized by the cross-flow velocity reaching zero in the center. It should also be noted that the swirl component of velocity is much greater than the radial-inward component. The feeding sheet (Figure 4.3) shows a large change in the magnitude of the cross-flow velocity across a very small region, with the vectors remaining parallel through the sheet. The cross-flow shock (Figure 4.4) is characterized by a change in both magnitude and direction of the vectors.

The angularity of the flow is presented in contour plots similar to that in Figure 4.5.

4.1. TOPOLOGICAL ASPECTS OF THE SOLUTIONS

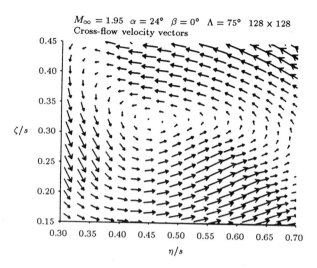

Figure 4.2: Cross-flow velocity vectors — Core region

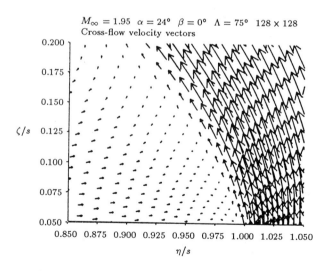

Figure 4.3: Cross-flow velocity vectors — Sheet region

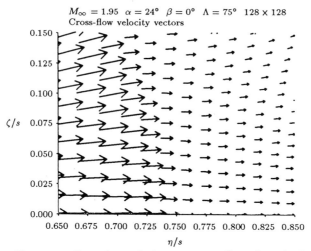

Figure 4.4: Cross-flow velocity vectors — Cross-flow shock region

The angularity is the deflection of the flow from the x-axis, in degrees. It is given by

$$\delta = \tan^{-1} \sqrt{\frac{v^2 + w^2}{u^2}}. \tag{4.1}$$

The angularity is zero on the symmetry plane, and changes very slowly on the windward side of the wing. There is a high gradient at the leading-edge, due to the rapid expansion. The angularity is high above and below the vortex, and decreases to a minimum in the center of vortex. Swirl angles of greater than 45° are reached in many of the cases in this book.

The cross-flow streamline plots can give a great deal of insight into the topology of the flow, but can also be very misleading. They are lines which are everywhere parallel to the cross-flow velocity. They are generated by a trajectory integration of the cross-flow velocities in the computational plane. The integration is described in Appendix B. This procedure yields cross-flow streamline plots similar to those shown in Figures 4.6 and 4.7. Figure 4.6 shows the cross-flow streamlines for the low Mach number and angle of attack case. The spiral structure of the vortex is clearly shown, and the vortex is almost circular in shape. Each cross-flow streamline converges into either the windward symmetry point (node), the leeward symmetry point (node), or the vortex (spiral). The attachment lines (saddle points) on the windward and leeward sides of the wing are also clear. Figure 4.7 shows the cross-flow streamlines for the case with higher values of M_∞ and α. The vortex is almost triangular in shape, and highly curved due to a strong cross-

4.2. PRESSURE PLOTS

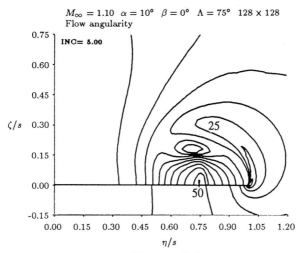

Figure 4.5: Flow angularity contours

flow shock under the vortex. Due to truncation errors in the trajectory integration and the extremely low radial velocities in the core, the trajectory integration reaches a limit cycle in the vortex in this case. This is one way in which the cross-flow streamline plots may be misleading. Another way is that the radial component of velocity may be much larger than the cross-flow components. Thus, where the cross-flow streamlines suggest flow towards the symmetry plane, the three-dimensional flow is really almost parallel to it. This ambiguity can be cleared up by looking at the tuft plots.

The tuft patterns show the flow direction at points on the leeward side of the wing. They are generated by interpolation of the computed velocities in one plane, and extended conically along the wing. Figure 4.8 shows the tuft pattern for the low Mach number case of Figure 4.6. As can be seen, the flow is highly turned under the vortex, but elsewhere it is almost entirely axial. While the cross-flow streamline pattern (Figure 4.6) suggests flow towards the symmetry plane, the tuft pattern shows that the cross-flow velocities there are much smaller than the radial component of velocity.

4.2 Pressure Plots

The variables considered here are the static pressure, total pressure and pitot pressure. The static pressure is shown through contour plots and line plots of the static pressure

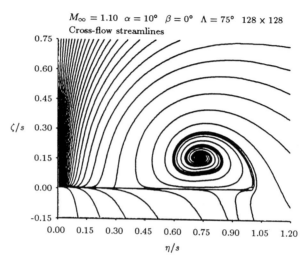

Figure 4.6: Cross-flow streamlines — $M_\infty = 1.1$, $\alpha = 10°$, $\Lambda = 75°$

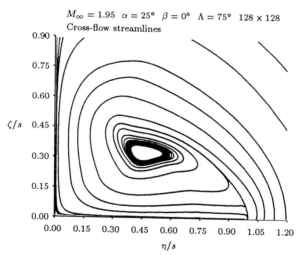

Figure 4.7: Cross-flow streamlines — $M_\infty = 1.95$, $\alpha = 25°$, $\Lambda = 75°$

4.2. PRESSURE PLOTS

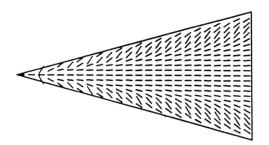

Figure 4.8: Tuft patterns

coefficient

$$C_p = \frac{p - p_\infty}{\frac{1}{2} p_\infty M_\infty^2}. \qquad (4.2)$$

Figure 4.9 is an example of the pressure coefficient on the wing. The leeward side pressure coefficient is characterized by a rapid expansion under the vortex, with recompression through a reverse cross-flow shock. The pressure inboard of the vortex is nearly constant. The entire windward side is at almost a constant pressure. Contours of the pressure coefficient (See Figure 4.10) show the low pressure region in the core and the rapid expansion at the leading-edge. Again, constant pressure regions on the windward side and on the leeward side, inboard of the vortex, are shown.

The total pressure is shown by plots of the total pressure coefficient

$$C_{p_0} = \frac{p_{0_\infty} - p_0}{p_{0_\infty}}. \qquad (4.3)$$

Figure 4.11 is a plot of the total pressure coefficient on the wing for the high Mach number and angle of attack case. The windward side of the wing has a constant total pressure coefficient, equal to the loss through the windward symmetry point on the bow shock. There is a large loss at the leading-edge (with a spurious overshoot). The leeward side has a constant total pressure inboard of the vortex, equal to the loss through the cross-flow shock inboard of the vortex. There is a further loss through the cross-flow shock under the vortex (with a spurious overshoot) and a constant value outboard of the shock. The

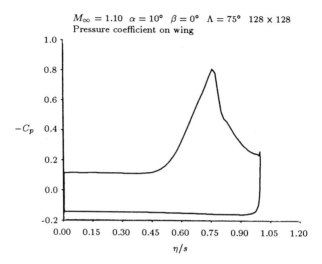

Figure 4.9: Pressure coefficient on wing

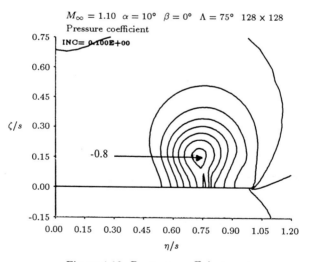

Figure 4.10: Pressure coefficient contours

4.3. MACH NUMBER AND DENSITY

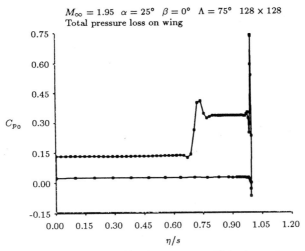

Figure 4.11: Total pressure coefficient on wing

contour plot of total pressure coefficient, shown in Figure 4.12, shows large losses in the feeding sheet and core, losses through the cross-flow shocks, and zero loss elsewhere.

The pitot pressure is what would be measured by a probe inserted in the flow. The pitot pressure is not equal to the local stagnation pressure, due to the fact that the probe generates a bow shock. It is computed from the formula

$$p_p = p_{0_\infty} \left[\frac{2\gamma}{\gamma+1} M^2 - \frac{\gamma-1}{\gamma+1} \right]^{\frac{-1}{\gamma-1}} \left[\frac{\frac{\gamma+1}{2} M^2}{1 + \frac{\gamma-1}{2} M^2} \right]^{\frac{\gamma}{\gamma-1}}, \quad (4.4)$$

which assumes that the probe is aligned with the local flow direction (a difficult feat in a vortex flow experiment). At angularities of greater than 20°, this formula will not give a good approximation to the measured pitot pressure [30]. The pitot pressure is normalized by the freestream total pressure for the plots. Contours of the pitot pressure for the high Mach number and angle of attack case are shown in Figure 4.13. The pitot pressure reaches a minimum in the core.

4.3 Mach Number and Density

The variables presented here are the Mach number, its cross-flow and radial components, and the density. All plots in this section are for the low Mach number and angle of attack case. Figure 4.14 shows contours of Mach number for this case. The Mach number is

52   CHAPTER 4. BASIC CHARACTERISTICS OF SOLUTIONS

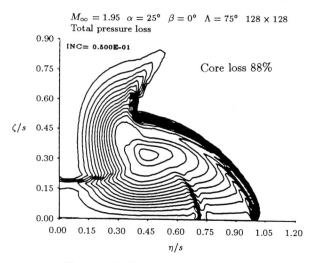

Figure 4.12: Total pressure coefficient contours

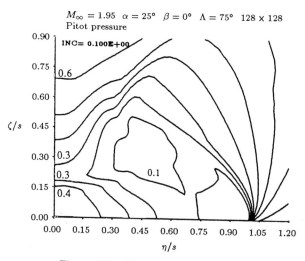

Figure 4.13: Pitot pressure coefficient contours

4.3. MACH NUMBER AND DENSITY

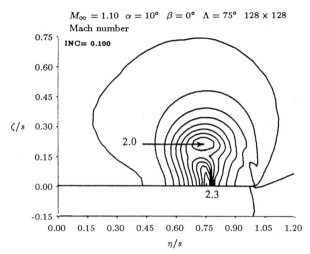

Figure 4.14: Mach number contours

nearly constant on the windward side, increasing through the expansion at the leading-edge. The Mach number reaches its maximum values above and below the vortex. The deceleration through the cross-flow shock is shown. The radial Mach number is given by

$$M_r = \sqrt{\frac{1}{c^2} \frac{(u + \eta v + \zeta w)^2}{1 + \eta^2 + \zeta^2}}. \qquad (4.5)$$

It is nearly constant outside the vortex, and reaches its maximum value in the core (See Figure 4.15). The cross-flow Mach number, given by

$$M_{cf} = \sqrt{M^2 - M_r^2}, \qquad (4.6)$$

(See Figure 4.16) is nearly constant on the windward side, but changes rapidly through the leading-edge expansion and the feeding sheet of the vortex. It reaches it maximum values above and below the vortex. The gradients are highest at the cross-flow shock and in the core, where the cross-flow Mach number reaches zero.

The contours of density, shown in Figure 4.17, reach a minimum in the core of the vortex. The density is nearly constant on the windward side, and on the leeward side inboard of the vortex. The expansion fan at the leading-edge and the expansion and recompression under the vortex are shown.

CHAPTER 4. BASIC CHARACTERISTICS OF SOLUTIONS

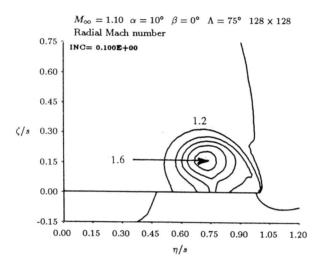

Figure 4.15: Radial Mach number contours

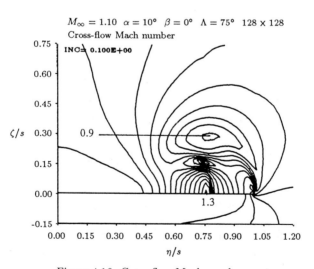

Figure 4.16: Cross-flow Mach number contours

4.4. VORTICITY   55

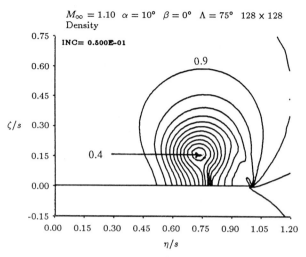

Figure 4.17: Density contours

4.4 Vorticity

The expression for the vorticity is

$$\omega = \frac{1}{x}\left[\left(\frac{\partial w}{\partial \eta}-\frac{\partial v}{\partial \zeta}\right)\hat{\mathbf{e}}_x + \left(\frac{\partial u}{\partial \zeta}+\eta\frac{\partial w}{\partial \eta}+\zeta\frac{\partial w}{\partial \zeta}\right)\hat{\mathbf{e}}_y - \left(\eta\frac{\partial v}{\partial \eta}+\zeta\frac{\partial v}{\partial \zeta}+\frac{\partial u}{\partial \eta}\right)\hat{\mathbf{e}}_z\right]. \quad (4.7)$$

Figure 4.18 shows contours of the log of the magnitude of the vorticity,

$$\Omega = \log\left(1 + \sqrt{\omega_x^2 + \omega_y^2 + \omega_z^2}\right). \quad (4.8)$$

The log function is used because of the extremely high vorticity at the leading-edge. Thus only quantitative information may be gained from the vorticity plots. The high gradients at the leading-edge and the sheet are clearly shown, along with the gradients at the cross-flow shocks.

4.5 Grid Resolution Effects

Line and contour plots of the cross-flow Mach number are shown for the low Mach number case on three different grids in Figures 4.19-4.24. The differences between the solution on the 64×64 grid (Figures 4.19 and 4.20) and the solution on the 128×128 grid (Figures 4.21 and 4.22) are substantial: on the 64 × 64 grid the cross-flow Mach number reaches a value of 0.7 above and below the vortex; on the 128 × 128 grid it reaches 0.9 above the vortex

56 CHAPTER 4. BASIC CHARACTERISTICS OF SOLUTIONS

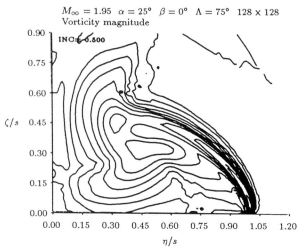

Figure 4.18: Vorticity contours

Table 4.1: Effects of grid refinement on normal force

Grid resolution	Normal Force Coefficient
64×64	0.4003
128×128	0.4087
256×256	0.4044

and 1.3 below it. The solution on the 256×256 grid (Figures 4.23 and 4.24) is very similar to that on the 128×128 grid. The vortex core, feeding-sheet and cross-flow shock are better defined on the finer mesh, but the maximum levels of cross-flow Mach number remain the same. Other flow variables (with the exception of total pressure, to be discussed in Chapter 5) show similar behavior. The normal force coefficients for different grid refinements are shown in Table 4.5. It is relatively insensitive to grid refinement. The normal forces calculated by the conical Euler method compare well with experimentally measured values [50]. The base grids chosen for the symmetric flat plate cases in this book have 128×128 equivalent global refinement; the vortex flap cases are run on 256×256 equivalent global refinement grids to resolve the hinge-line region.

4.5. GRID RESOLUTION EFFECTS

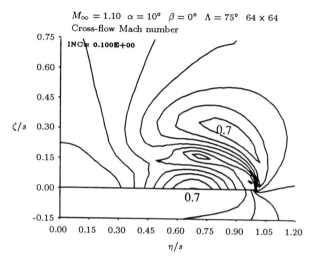

Figure 4.19: Cross-flow Mach number contours — 64×64 grid

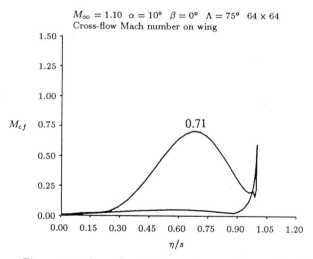

Figure 4.20: Cross-flow Mach number on wing — 64×64 grid

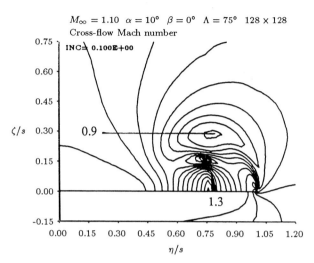

Figure 4.21: Cross-flow Mach number contours — 128×128 grid

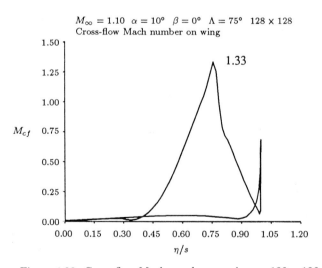

Figure 4.22: Cross-flow Mach number on wing — 128×128 grid

4.5. GRID RESOLUTION EFFECTS

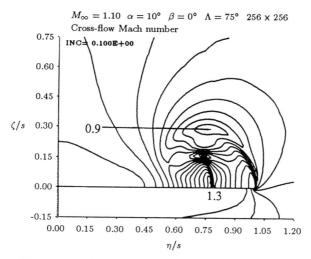

Figure 4.23: Cross-flow Mach number contours — 256×256 grid

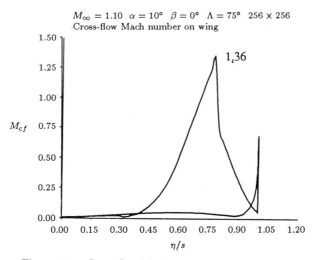

Figure 4.24: Cross-flow Mach number on wing — 256×256 grid

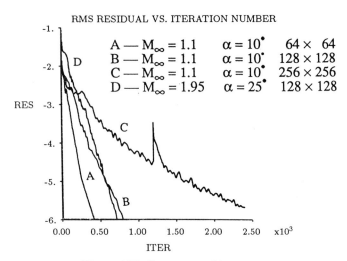

Figure 4.25: Convergence history

4.6 Convergence

A convergence history for several cases is shown in Figure 4.25. The jump in the 256 × 256 case residual at 1200 iterations is due to the fact that the case was restarted. All the cases presented in this book were converged at least four orders of magnitude. The convergence parameter used was the L_2 norm of the density residual. All of the 128 × 128 equivalent global refinement cases converged in less than 1000 iterations, and all of the 256 × 256 cases in less than 2500 iterations.

Another test of convergence is given by the total enthalpy. In a converged solution the total enthalpy should be everywhere constant. Figure 4.26 is a plot of the total enthalpy loss $1 - h_0/h_{0_\infty}$ on the wing for the low Mach number and angle of attack case. As can be seen, the total enthalpy is constant to within a thousandth of a percent. All the cases in this book have total enthalpy constant to within a hundredth of a percent.

Computations were carried out on an Alliant FX/8-3, a three-processor machine with thirty-two element vector registers. The code was fully vectorized, so that all loops in the flux and damping calculations ran in vector-concurrent mode. The code is presented in Appendix C. Computation times and speed-up due to vectorization are shown in Table 4.6. Alliant compiler directive statements (source lines beginning with CVD$) were used to

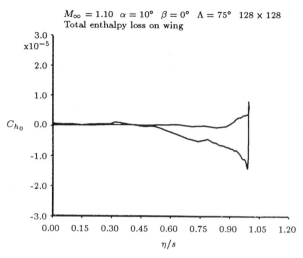

Figure 4.26: Total enthalpy coefficient on wing

Table 4.2: Vectorization and concurrency speed-ups

Optimization	Iterations/node/second
None	78
Global	218
Global, vector	515
Global, concurrent	648
Global, vector, concurrent	1275

aid in the vectorization process.

Large savings were achieved by the use of embedded regions. For the 256 × 256 grid case presented in Chapter 5, for instance, the present scheme required 6500 nodes, 2500 iterations and 4.0×10^5 words of memory. On a structured mesh of 256 × 256 refinement everywhere, 66,000 nodes, 8000 iterations (estimated) and 2.6×10^6 words of memory would be required. This represents a savings of almost an order of magnitude in storage, and more than an order of magnitude in computation time.

4.7 Summary

This chapter has described the output from the solution scheme and how the features of the flow are manifested in the various output flow variables. The topology of the flow is described *via* cross-flow velocity vectors, cross-flow streamlines, flow angularity contours and tuft pattern plots. The static, total and pitot pressures show the leading-edge expansion, the vortex and cross-flow shocks. Mach number, cross-flow Mach number and axial Mach number are also shown. The vorticity is presented *via* plots of the log of the magnitude of the vorticity. The solutions have been shown to be mesh-converged on an equivalent 128×128 grid. Two different convergence criteria for the scheme have been described. The savings achieved by using embedded regions has been outlined.

Chapter 5

Total Pressure Losses — A A Numerical Study

Numerical solutions of the Euler and Navier-Stokes equations exhibit large total pressure losses which are localized to the vicinity of the vortex and its feeding sheet [1]. Computational experience has shown that the level of these losses is nearly independent of computational parameters such as mesh spacing, artificial viscosity level and differencing scheme. They are dependent upon the physical parameters of the flow — the free-stream Mach number, the angles of attack and yaw and the leading-edge sweep of the wing.

This chapter will demonstrate this independence of numerical parameters and dependence upon physical parameters. A comparison of the losses with those seen in experiment will be presented. A study will be made of the level of artificial viscosity in the vortex, both by estimating the equivalent Reynolds number in the core, and by examining the level to which Crocco's relation is satisfied. Lossless solutions, computed from an altered set of equations, will be presented and compared with solutions produced by the solution method presented in Chapter 3.

---

[1]This refers to schemes in which entropy is not treated as a state variable. Schemes which enforce $DS/Dt = 0$ yield solutions in which the only losses are those introduced at shocks which are "fit" into the flow-field [34].

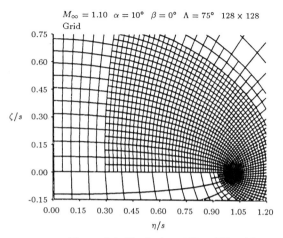

Figure 5.1: Equivalent 128 × 128 grid

5.1 Effects of Computational Parameters

The level of the total pressure loss predicted by computations has been found to be basically independent of the damping levels and grid refinement. Results are presented here for a 75° swept wing at an angle of attack of 10° in a Mach number 1.1 flow. This case is run at different levels of damping and mesh refinement, and the loss levels are shown. The distribution of losses changes, but the level of the loss remains very nearly constant over a wide range of damping levels and grid refinements.

The first result presented is for

- Equivalent 128 × 128 refinement ,

- Second-difference smoothing coefficient $\epsilon_2 = 0.003$,

- Fourth-difference smoothing coefficient $\epsilon_4 = 0.001$.

This case was described in Chapter 4, but the features of the flow are summarized here. The grid, shown in Figure 5.1, is doubly embedded, with an equivalent refinement in the region of the vortex of a 128 × 128 global grid. The cross-flow streamlines are shown in Figure 5.2. The notable features in this flow are the nodes at the windward and leeward symmetry lines on the wing, the saddle points on the windward and leeward sides of the wing, and the vortex. Contours of the pressure coefficient are shown in Figure 5.3.

5.1. EFFECTS OF COMPUTATIONAL PARAMETERS

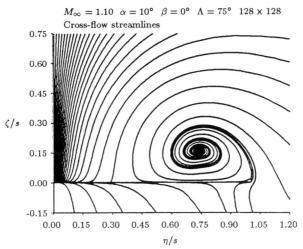

Figure 5.2: Cross-flow streamlines

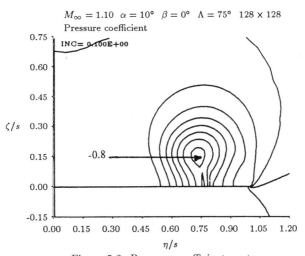

Figure 5.3: Pressure coefficient contours

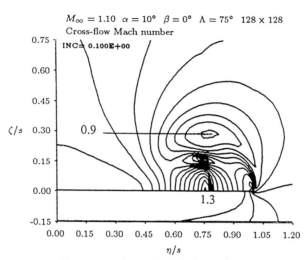

Figure 5.4: Cross-flow Mach number contours

5.1. EFFECTS OF COMPUTATIONAL PARAMETERS

The low pressure region in the vortex core and underneath it, on the wing, are clearly shown. Cross-flow Mach number contours are shown in Figure 5.4. The flow expands to a cross-flow Mach number of 0.9 above the vortex, and 1.3 underneath it, leading to a reverse cross-flow shock under the vortex.

The contours of total pressure loss are shown in Figure 5.5. The loss across the bow shock is negligible (less that 0.1%) for this case. Although the bow shock loss is larger than this in other cases presented in this book, it remains negligible with respect to the losses due to vortices and cross-flow shocks in each case. The loss due to the cross-flow shock is on the order of 2%. As can be seen, the level of the loss in the core is 37%.

In order to study the effects of damping level, the following cases are presented, each on a grid with the equivalent of 128×128 refinement.

1. $\epsilon_2 = .003$, $\epsilon_4 = .0003$,

2. $\epsilon_2 = .003$, $\epsilon_4 = .003$,

3. $\epsilon_2 = .001$, $\epsilon_4 = .001$,

4. $\epsilon_2 = .010$, $\epsilon_4 = .001$.

The second-difference damping coefficients are approximately an order of magnitude lower than are typically used. This is possible due to the fact that the cross-flow in this case is only weakly transonic. With these low values of ϵ_2, the fourth-difference and second-difference damping are of similar magnitudes. In a typical case, with higher M_∞, the second-difference damping effects will dominate.

The total pressure loss contours for these cases are shown in Figures 5.6-5.9. The loss in the feeding sheet changes with the damping, dropping below the 2% level in Figure 5.7. The level of the core total pressure loss is basically insensitive to an order of magnitude change in either of the damping coefficients, however. The location of the center of the core is also remarkably insensitive to these changes; the point of maximum total pressure loss is in the same cell in each of the cases. The distribution of the loss changes dramatically, however. The loss levels and two scales for the sizes of the vortices in the different cases are tabulated in Table 5.1. "Core sizes" are determined by summing the area of all cells in which $1 - p_0/p_{0\infty} \geq 10\%$. They are normalized by the core size for the $\epsilon_2 = 0.003$, $\epsilon_4 = 0.001$ case. "Vortex width" and "vortex height" are determined by measuring the

68 CHAPTER 5. TOTAL PRESSURE LOSSES — A NUMERICAL STUDY

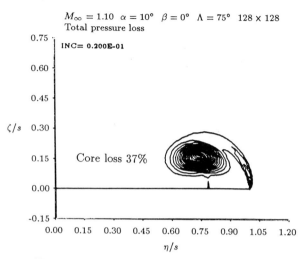

Figure 5.5: Total pressure loss — $\epsilon_2 = 0.003$ $\epsilon_4 = 0.001$

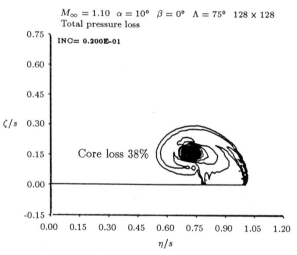

Figure 5.6: Total pressure loss — $\epsilon_2 = 0.003$ $\epsilon_4 = 0.0003$

5.1. EFFECTS OF COMPUTATIONAL PARAMETERS

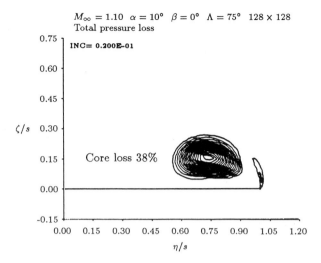

Figure 5.7: Total pressure loss — $\epsilon_2 = 0.003$ $\epsilon_4 = 0.003$

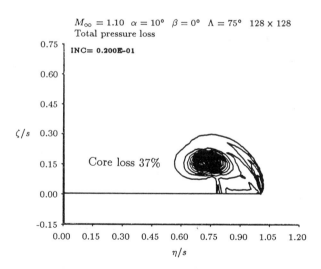

Figure 5.8: Total pressure loss — $\epsilon_2 = 0.001$ $\epsilon_4 = 0.001$

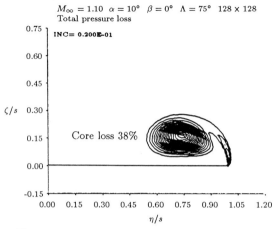

Figure 5.9: Total pressure loss — $\epsilon_2 = 0.010$ $\epsilon_4 = 0.001$

Table 5.1: Effects of artificial viscosity level on loss

ϵ_2	ϵ_4	Loss Level	Core Size	Vortex Width	Vortex Height
0.003	0.0010	37.4%	1.00	1.00	1.00
0.003	0.0003	37.6%	0.32	1.10	1.13
0.003	0.0030	38.2%	1.74	1.08	1.02
0.001	0.0010	37.2%	0.71	0.97	1.01
0.010	0.0010	37.6%	1.67	1.11	1.08

width and height of the 2% contour at the center of the vortex. They are also normalized by the values of the $\epsilon_2 = 0.003$, $\epsilon_4 = 0.001$ case. The "core size", a microscale for the vortex, changes dramatically with the damping coefficients. The "vortex width" and "vortex height", however, are much less sensitive.

The level of artificial viscosity does not greatly affect flow variables outside the core. Contours of pressure coefficient and cross-flow Mach number for the $\epsilon_2 = 0.003$, $\epsilon_4 = 0.0003$ case are shown in Figures 5.10 and 5.11. Comparison with Figures 5.3 and 5.4 shows a lower pressure in the vortex core, and a smaller region of inflection for the cross-flow Mach number.

In order to study the effects of grid refinement, the following cases are presented, each with damping coefficients $\epsilon_2 = 0.01$ and $\epsilon_4 = 0.001$.

1. Equivalent 64 × 64 resolution ,

5.1. EFFECTS OF COMPUTATIONAL PARAMETERS

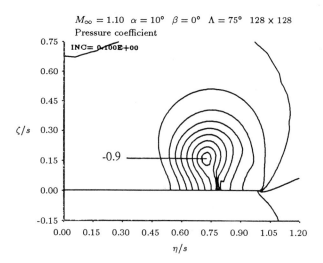

Figure 5.10: Pressure Coefficient — $\epsilon_2 = 0.003$ $\epsilon_4 = 0.0003$

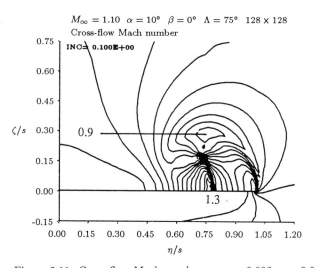

Figure 5.11: Cross-flow Mach number — $\epsilon_2 = 0.003$ $\epsilon_4 = 0.0003$

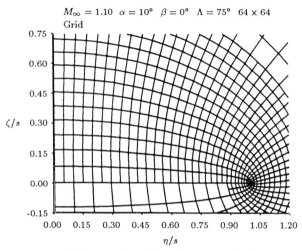

Figure 5.12: Grid — 64 × 64 resolution

Table 5.2: Effects of grid refinement on loss

Equivalent grid	Loss Level	Core Size	Vortex Width	Vortex Height
64 × 64	33.9%	2.55	1.39	1.43
128 × 128	37.6%	1.00	1.00	1.00
256 × 256	35.1%	0.26	0.96	1.05

2. Equivalent 128 × 128 resolution,

3. Equivalent 256 × 256 resolution.

The grid, pressure coefficient contours, cross-flow Mach number contours and total pressure loss contours for the case with equivalent 64×64 resolution are shown in Figures 5.12-5.15. They are shown for the 256 × 256 case in Figures 5.16-5.19. The distributions of cross-flow Mach number and pressure coefficient change drastically from the 64 × 64 grid to the 128 × 128 grid, less so from the 128 × 128 grid to the 256 × 256 grid. The level of core total pressure loss, however, is relatively insensitive. The distribution changes with refinement, with the vortex core size diminishing as the grid is refined. The core loss level and core sizes on the different grids are given in Table 5.1. On the finest grid, the spiral structure of the feeding sheet may be seen.

It is interesting to note that, despite the independence of the level of the loss on the level of the artificial viscosity, the level of the loss does depend on the *form* of the viscosity.

5.1. EFFECTS OF COMPUTATIONAL PARAMETERS

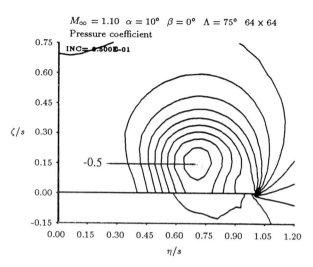

Figure 5.13: Pressure Coefficient — 64 × 64 resolution

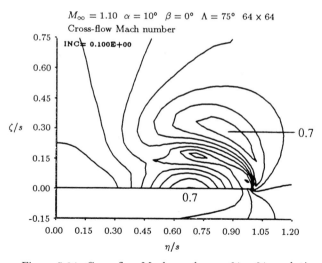

Figure 5.14: Cross-flow Mach number — 64 × 64 resolution

74   CHAPTER 5. TOTAL PRESSURE LOSSES — A NUMERICAL STUDY

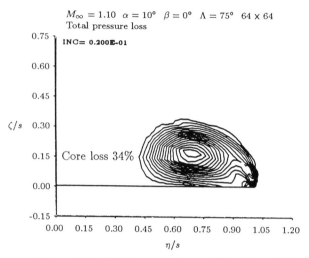

Figure 5.15: Total pressure loss — 64 × 64 resolution

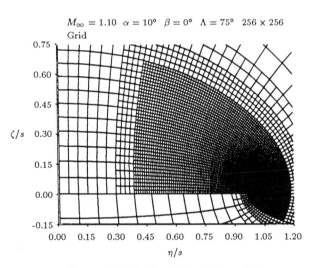

Figure 5.16: Grid — 256 × 256 resolution

5.1. EFFECTS OF COMPUTATIONAL PARAMETERS

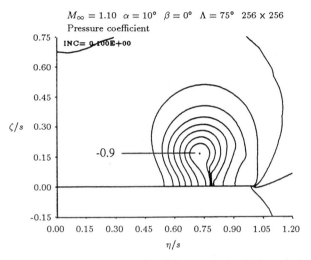

Figure 5.17: Pressure Coefficient — 256 × 256 resolution

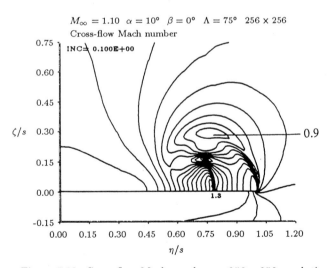

Figure 5.18: Cross-flow Mach number — 256 × 256 resolution

76   CHAPTER 5.   TOTAL PRESSURE LOSSES — A NUMERICAL STUDY

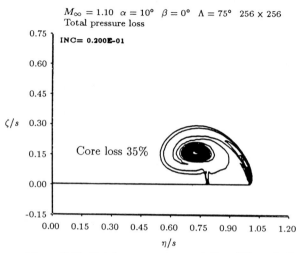

Figure 5.19: Total pressure loss — 256 × 256 resolution

This dependence on the form of the artificial viscosity was discovered in an attempt to lower the damping on the continuity equation. Figure 5.20 shows the total pressure loss contours for the $M_\infty = 1.1$, $\alpha = 10°$, $\Lambda = 75°$ case on an equivalent 128 × 128 grid where $\epsilon_2 = 0.010$ and $\epsilon_4 = 0.001$, but the second-difference damping terms on the continuity equation have been removed. For this case, the loss in the core is 27%.

The effect that zeroing the continuity damping has on the form of the damping for the system of equations may be seen most easily by analyzing a one-dimensional steady model problem. The equivalent continuity and momentum equations are, including the second-difference damping,

$$\frac{\partial}{\partial x}(\rho u) = \epsilon_\rho \frac{\partial^2 \rho}{\partial x^2}, \tag{5.1a}$$

$$\frac{\partial}{\partial x}\left(\rho u^2 + p\right) = \epsilon_{\rho u} \frac{\partial^2}{\partial x^2}(\rho u), \tag{5.1b}$$

where ϵ_ρ and $\epsilon_{\rho u}$ are damping coefficients for the two equations. Using the chain rule, the momentum equation may be written

$$\rho u \frac{\partial u}{\partial x} + u \frac{\partial}{\partial x}(\rho u) + \frac{\partial p}{\partial x} = \epsilon_{\rho u} \left(u \frac{\partial^2 \rho}{\partial x^2} + \rho \frac{\partial^2 u}{\partial x^2} + 2 \frac{\partial \rho}{\partial x} \frac{\partial u}{\partial x} \right). \tag{5.2}$$

which, when combined with continuity, becomes

$$\rho u \frac{\partial u}{\partial x} + \frac{\partial p}{\partial x} = \rho \epsilon_{\rho u} \frac{\partial^2 u}{\partial x^2} + u \left(\epsilon_{\rho u} - \epsilon_\rho \right) \frac{\partial^2 \rho}{\partial x^2} + 2 \epsilon_{\rho u} \frac{\partial \rho}{\partial x} \frac{\partial u}{\partial x}. \tag{5.3}$$

5.1. EFFECTS OF COMPUTATIONAL PARAMETERS

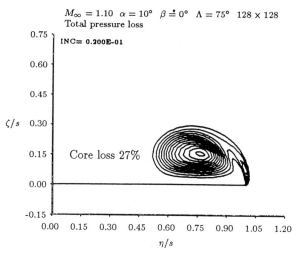

Figure 5.20: Total pressure loss — Zero second-difference continuity damping

Comparison with the one-dimensional steady Navier-Stokes equation shows that this equation has two spurious terms: the term $u(\epsilon_{\rho u} - \epsilon_\rho)\partial^2 \rho/\partial x^2$, due to the difference in the two damping coefficients, and the cross term $2\epsilon_{\rho u}\,\partial \rho/\partial x\,\partial u/\partial x$.

If $\epsilon_{\rho u} = \epsilon_\rho$, the first spurious term drops out. This suggests a way to zero the continuity damping while retaining the proper form of damping for the system of equations. By subtracting this term from the momentum and energy equations, the system will have the same form as it would in the case $\epsilon_{\rho u} = \epsilon_\rho$. For the conical Euler equations, this term has the form

$$\delta_{\rho u} = u\left(\epsilon_{\rho u} - \epsilon_\rho\right) D_2(\rho), \tag{5.4a}$$

$$\delta_{\rho v} = v\left(\epsilon_{\rho v} - \epsilon_\rho\right) D_2(\rho), \tag{5.4b}$$

$$\delta_{\rho w} = w\left(\epsilon_{\rho w} - \epsilon_\rho\right) D_2(\rho), \tag{5.4c}$$

$$\delta_{\rho E} = h_0\left(\epsilon_{\rho E} - \epsilon_\rho\right) D_2(\rho), \tag{5.4d}$$

where D_2 is the second-difference damping operator defined in Chapter 3. Figure 5.21 shows the total pressure loss contours for same case as Figure 5.20, with zero second-difference damping on the continuity equation and the spurious terms due to the difference in damping coefficients subtracted from the remaining equations. The core total pressure loss is 38%, as it was in the case without lowering the continuity equation damping.

78 CHAPTER 5. TOTAL PRESSURE LOSSES — A NUMERICAL STUDY

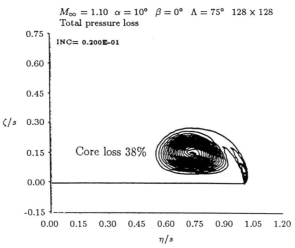

Figure 5.21: Total pressure loss — Zero second-difference continuity damping

The damping retains the spurious cross term, $2\epsilon_{\rho u}\,\partial\rho/\partial x\,\partial u/\partial x$, so that the damped Euler equations differ in form from the Navier-Stokes equations. In regions where the product of the cross term is much smaller than the second-derivative term, that is

$$\left|2\frac{\partial\rho}{\partial x}\frac{\partial u}{\partial x}\right| \ll \left|\rho\frac{\partial^2 u}{\partial x^2}\right|, \tag{5.5}$$

the damped momentum equations will resemble the Navier-Stokes equations.

5.2 Effects of Physical Parameters

While the level of the total pressure loss is independent of computational parameters, there is a definite dependence upon the free-stream Mach number, the angle of attack, and the sweep of the wing. Table 5.2 gives the core loss level as a function of these three parameters, where each case has been run on a 128×128 grid. For conical flows there is evidence that the three-parameter space $(M_\infty, \alpha, \Lambda)$ may be collapsed to a two-parameter space (α_N, M_N) where

$$\alpha_N = \tan^{-1}\left(\frac{\tan\alpha}{\cos\Lambda}\right), \tag{5.6a}$$

$$M_N = M_\infty \cos\Lambda\sqrt{1 + \sin^2\alpha\tan^2\Lambda} \tag{5.6b}$$

[59]. While much of the experimental data has been correlated in this way, no explicit experiments have been done to verify that identical flow fields are produced at particular

Table 5.3: Effects of physical parameters on loss

M_∞	α	Λ	M_N	α_N	Loss level
1.10	10°	75.0°	0.28	34.3°	38%
1.70	8°	75.0°	0.44	28.5°	50%
1.70	12°	75.0°	0.44	39.4°	58%
1.95	10°	75.0°	0.50	34.3°	60%
1.70	12°	67.5°	0.65	29.0°	68%
1.95	25°	75.0°	0.51	61.0°	89%
2.00	20°	67.5°	0.77	43.6°	90%
2.40	20°	75.0°	0.63	54.6°	90%
2.80	20°	75.0°	0.73	54.6°	96%

coordinates α_N, M_N [62]. The values of α_N and M_N for each of the cases are also given in Table 5.2.

5.3 Comparison of Loss with Experiment

The primary basis for this section is the experiment of Monnerie and Werlé of a 75° swept delta wing at 10° and 25° angle of attack in a Mach number 1.95 flow [37]. The computed solution for the 10° case is shown in Figures 5.22-5.28. Figure 5.22 is the grid for the case, giving an equivalent 128 × 128 refinement. The cross-flow streamlines (Figure 5.23) show the topology of the flow. There are saddle points on the windward and leeward sides of the wing. All the cross-flow streamlines between the two saddle points converge into the large primary vortex. The other streamlines converge into the nodes at the windward and leeward symmetry points. The primary vortex shows the reverse curvature that is characteristic of a vortex with a strong cross-flow shock underneath it. The pressure coefficient, shown in Figure 5.24, shows the rapid expansion at the leading-edge and the low pressure region in the core. The Mach number (Figure 5.25) reaches a value of 3.4 above the vortex and 4.4 beneath it. The radial Mach number (Figure 5.26) reaches a value of 2.6 in the core and 2.8 beneath it. The cross-flow Mach number, shown in Figure 5.27, has very high gradient regions at the sheet, the core and the shock. It reaches a value of 1.5 above the vortex and 1.7 just upstream of the cross-flow shock beneath the vortex. Figure 5.28 shows the total pressure loss contours for this case, with 60% of the total pressure lost in the core.

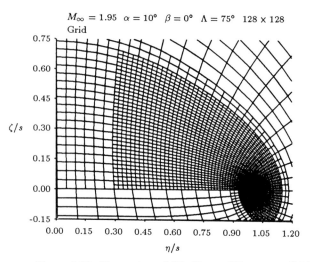

Figure 5.22: Monnerie and Werlé $\alpha = 10°$ case — Grid

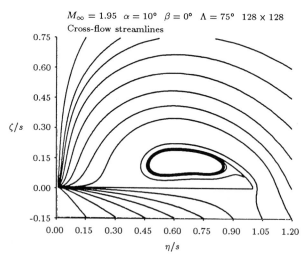

Figure 5.23: Monnerie and Werlé $\alpha = 10°$ case — Cross-flow streamlines

5.3. COMPARISON OF LOSS WITH EXPERIMENT

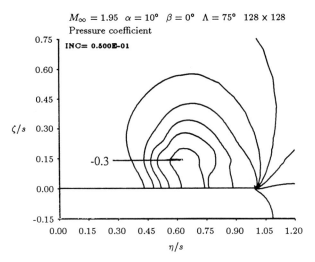

Figure 5.24: Monnerie and Werlé $\alpha = 10°$ case — Pressure coefficient

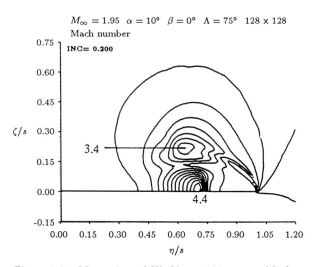

Figure 5.25: Monnerie and Werlé $\alpha = 10°$ case — Mach number

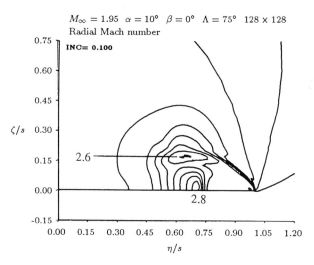

Figure 5.26: Monnerie and Werlé $\alpha = 10°$ case — Radial Mach number

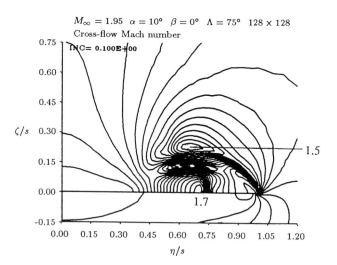

Figure 5.27: Monnerie and Werlé $\alpha = 10°$ case — Cross-flow Mach number

5.3. COMPARISON OF LOSS WITH EXPERIMENT

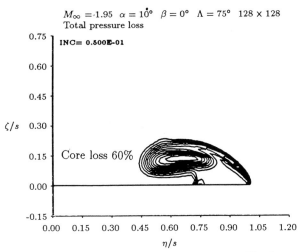

Figure 5.28: Monnerie and Werlé $\alpha = 10°$ case — Total pressure loss

The experimentally measured pitot pressures for this case are shown in Figure 5.29. Figure 5.30 shows the computed pitot pressures (defined in Chapter 4); Figure 5.31 shows the flow angularity in the computations.

The agreement is surprisingly good in the vortex. The boundary layer is obviously not modeled by the Euler equation calculations. The flow angularity is very high, particularly directly underneath the vortex. It is interesting to note that the contours of density, shown in Figure 5.32, are very similar to the contours of pitot pressure. This is due to the fact that, as the Mach number becomes large, the expression for the pitot pressure takes the form $p_p \sim \rho u_i u_i$ and $u_i u_i$ approaches its limiting value.

The $\alpha = 25°$ case has been described in Chapter 4. The computed pitot pressure (Figure 5.33) and the experimental pitot pressure (Figure 5.34) compare well. As with the $\alpha = 10°$ case, the density contours (Figure 5.35) bear a strong resemblance to the pitot pressure contours. A comparison of the measured and computed flow angularity on the leeward symmetry plane is given in Figure 5.36. The angularities agree well, particularly near the wing.

84   CHAPTER 5. TOTAL PRESSURE LOSSES — A NUMERICAL STUDY

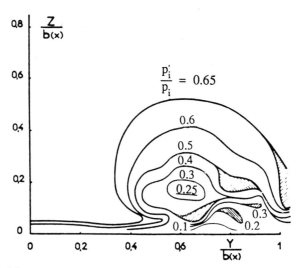

Figure 5.29: Monnerie and Werlé $\alpha = 10°$ case — Measured pitot pressure

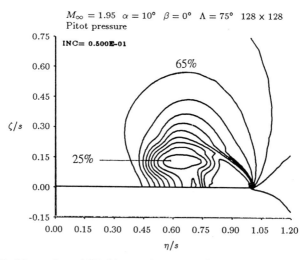

Figure 5.30: Monnerie and Werlé $\alpha = 10°$ case — Computational pitot pressures

5.3. COMPARISON OF LOSS WITH EXPERIMENT

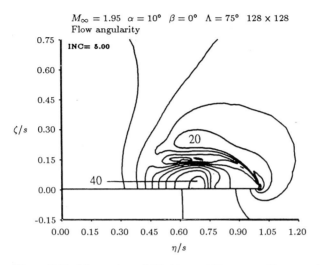

Figure 5.31: Monnerie and Werlé $\alpha = 10°$ case — Flow angularity

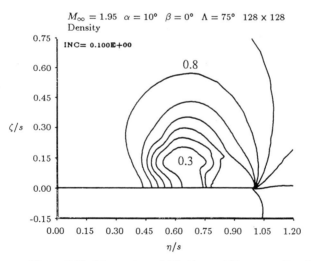

Figure 5.32: Monnerie and Werlé $\alpha = 10°$ case — Density

86   CHAPTER 5. TOTAL PRESSURE LOSSES — A NUMERICAL STUDY

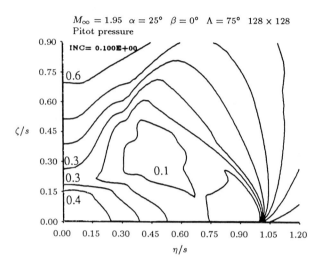

Figure 5.33: Monnerie and Werlé $\alpha = 25°$ case — Computational pitot pressures

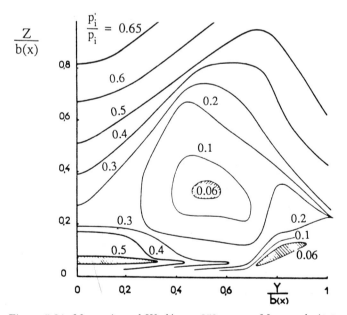

Figure 5.34: Monnerie and Werlé $\alpha = 25°$ case — Measured pitot pressure

5.3. COMPARISON OF LOSS WITH EXPERIMENT

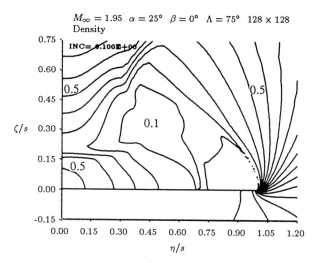

Figure 5.35: Monnerie and Werlé $\alpha = 25°$ case — Density

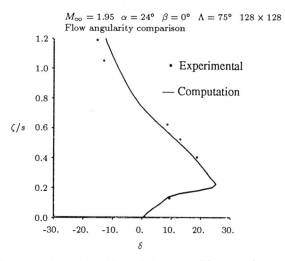

Figure 5.36: Monnerie and Werlé $\alpha = 25°$ case — Flow angularity comparison

5.4 Artificial Viscosity Level in the Vortex

In this section, the level of artificial viscosity in the vortex will be examined. Two approaches will be taken. In the first, the terms in Crocco's relation will be computed and compared, to ascertain the extent to which Crocco's relation is satisfied throughout the field. In the second, an equivalent Reynolds number will be calculated from the artificial viscosity coefficient, ϵ_2, and local values for the weighting function, δp, and the mesh spacing, Δ.

Crocco's relation comes from the definition of total enthalpy,

$$h_0 = \frac{u_i u_i}{2} + e + \frac{p}{\rho}, \tag{5.7}$$

the second law of thermodynamics,

$$T\frac{\partial S}{\partial x_i} = \frac{\partial e}{\partial x_i} + p\frac{\partial}{\partial x_i}\left(\frac{1}{\rho}\right), \tag{5.8}$$

the inviscid Cartesian momentum equations,

$$\frac{1}{\rho}\frac{\partial p}{\partial x_i} = -u_j\frac{\partial u_i}{\partial x_j}, \tag{5.9}$$

and the vector identity

$$u_j\frac{\partial u_i}{\partial x_j} = \frac{\partial}{\partial x_i}\left(\frac{u_j u_j}{2}\right) - \epsilon_{ijk}u_j\omega_k. \tag{5.10}$$

In the above equations, S is the entropy, T is the static temperature and ϵ_{ijk} is the alternate tensor, defined by

$$\epsilon_{ijk} = \begin{cases} 1 & \text{if } (i,j,k) \text{ are an even permutation of } (1,2,3); \\ -1 & \text{if } (i,j,k) \text{ are an odd permutation of } (1,2,3); \\ 0 & \text{otherwise.} \end{cases} \tag{5.11}$$

Combining these gives

$$\frac{\partial h_0}{\partial x_i} = T\frac{\partial S}{\partial x_i} + \epsilon_{ijk}u_j\omega_k. \tag{5.12}$$

Non-dimensionalizing S by C_v and all other variables as before, Crocco's relation becomes

$$\frac{\partial h_0}{\partial x_i} = \frac{1}{\gamma - 1}\frac{p}{\rho}\frac{\partial S}{\partial x_i} + \epsilon_{ijk}u_j\omega_k. \tag{5.13}$$

There are two ways of satisfying this relation in an isenthalpic vortical flow. One is that S is constant, so that the gradient of the entropy is zero. For this to be true, $\epsilon_{ijk}u_j\omega_k$ must be zero, that is, the vorticity vector must be parallel to the velocity vector. This is

5.4. ARTIFICIAL VISCOSITY LEVEL IN THE VORTEX

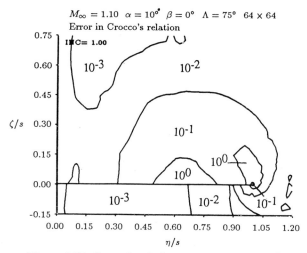

Figure 5.37: Crocco's relation error — 64 × 64 refinement

Beltrami flow. The other possibility is that the velocity vector and vorticity vector are *not* parallel, so that $\partial S/\partial x_i$ is non-zero.

The degree to which Crocco's relation is satisfied is a measure of the effects of artificial viscosity on the solutions. In regions where the artificial viscosity terms are negligible, Crocco's relation will be satisfied. In regions where they are not negligible, the terms in Crocco's relation will not balance. Figures 5.37-5.39 are contour plots of the log of the error, ε, in Crocco's relation where ε is given by

$$\varepsilon_i = \frac{\partial h_0}{\partial x_i} - \frac{1}{\gamma - 1}\frac{p}{\rho}\frac{\partial S}{\partial x_i} - \epsilon_{ijk}u_j\omega_k \ ,$$
$$\varepsilon = \sqrt{\varepsilon_i\varepsilon_i} \ . \tag{5.14}$$

Orders of magnitude are indicated on the plots. On each of the grids, the error in Crocco's relation is order one in the vortex and at the leading-edge, and smaller elsewhere.

An equivalent Reynolds number may be calculated by studying the form of the artificial viscosity. In regions in which the second difference viscosity dominates the fourth difference viscosity, the damping terms on the conical Euler equations have the form

$$\begin{aligned} D\left(\hat{\mathbf{U}}\right) &\sim \epsilon_2 \bar{L}\left(\hat{\mathbf{U}}, \delta p\right) \\ &\sim \epsilon_2 \delta p L\left(\hat{\mathbf{U}}\right) \\ &\sim 4\epsilon_2 \delta p \Delta^2 \nabla^2 \hat{\mathbf{U}}, \end{aligned} \tag{5.15}$$

90   CHAPTER 5. TOTAL PRESSURE LOSSES — A NUMERICAL STUDY

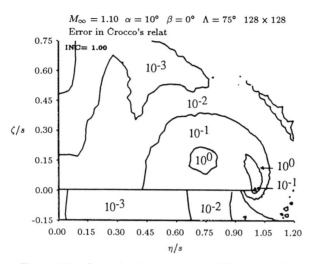

Figure 5.38: Crocco's relation error — 128×128 refinement

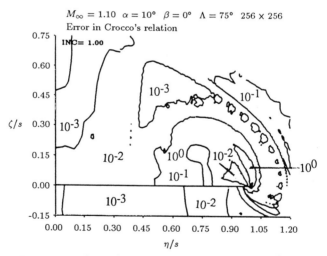

Figure 5.39: Crocco's relation error — 256×256 refinement

5.4. ARTIFICIAL VISCOSITY LEVEL IN THE VORTEX

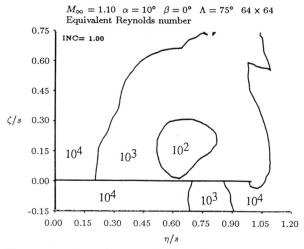

Figure 5.40: Equivalent Reynolds number — 64 × 64 refinement

where Δ is an estimate of the grid spacing, and the factor of four arises from the non-normalized stencil used for the Laplacian operator L. Since this damping operator is multiplied by the CFL number in the temporal updating, the actual damping terms added to the conical Euler equations have the form

$$\lambda D \left(\hat{\mathbf{U}} \right) \sim 4\lambda \epsilon_2 \delta p \Delta \nabla^2 \hat{\mathbf{U}}, \qquad (5.16)$$

where λ is the CFL number. Comparison with the Navier-Stokes equations suggests an "equivalent" Reynolds number, defined by

$$\frac{1}{Re_{eq}} = 4\lambda \epsilon_2 \delta p \Delta \quad , \qquad (5.17)$$

where Δ is taken to be the average of the lengths of the faces emanating from the node. Contours of the log of the equivalent Reynolds number are shown in Figures 5.40-5.42 for the same $M_\infty = 1.1$, $\alpha = 10°$, $\Lambda = 75°$ case with different grid refinements. Orders of magnitude of the equivalent Reynolds numbers in different regions are shown. The equivalent Reynolds number increases with grid refinement, reaching values on the order of 10^3 in the core for the 256 × 256 case. The equivalent Reynolds number is surprisingly high at the leading-edge, due to the small size of the cells.

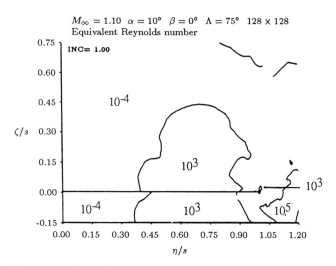

Figure 5.41: Equivalent Reynolds number — 128×128 refinement

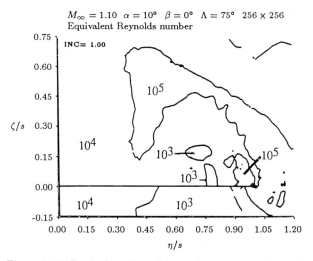

Figure 5.42: Equivalent Reynolds number — 256×256 refinement

5.4. ARTIFICIAL VISCOSITY LEVEL IN THE VORTEX

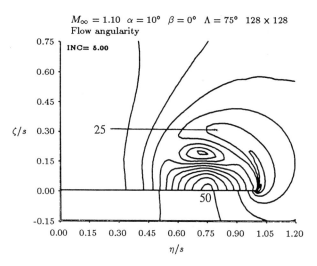

Figure 5.43: Isentropic case — Flow angularity

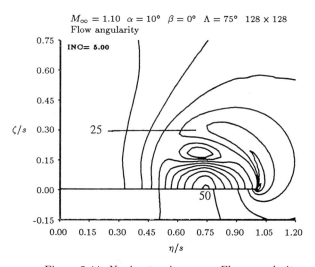

Figure 5.44: Nonisentropic case — Flow angularity

94 CHAPTER 5. TOTAL PRESSURE LOSSES — A NUMERICAL STUDY

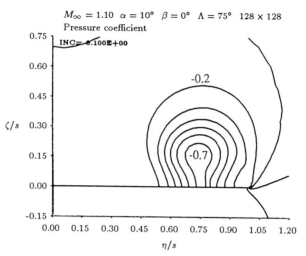

Figure 5.45: Isentropic case — Pressure coefficient

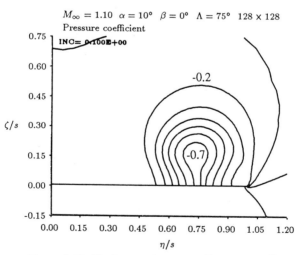

Figure 5.46: Nonisentropic case — Pressure coefficient

5.5 Lossless Solutions to the Euler Equations

It is possible to generate lossless flows by modifying the approach taken in this book. At each iteration, the streamwise momentum is changed to enforce constant total pressure. This is equivalent to replacing the streamwise momentum equation with

$$\frac{\partial p_0}{\partial s} = 0 \; , \qquad (5.18)$$

where s is a streamwise coordinate. This method will not apply in cases with strong shocks. A case has been calculated by this method to help understand the difference between the lossless case, in which the vorticity is locally aligned with the velocity, and the cases computed by the usual method. Results are shown here for the $M_\infty = 1.1$, $\alpha = 10°$, $\Lambda = 75°$ case with the isentropic and non-isentropic methods.

Figures 5.43 and 5.44 show the flow angularity for the two cases. In each case, the flow is turned 50° underneath the vortex and 25° above it. The only difference evident in the two plots is in the center of the core, and even there the difference is slight. Figures 5.45 and 5.46 show the pressure coefficient contours for the two cases. Again the differences are very slight. The Mach number, shown in Figures 5.47 and 5.48, does show differences. In the isentropic case, the maximum value of M occurs at the center of the vortex; in the non-isentropic case, maxima occur above and below the vortex. The Mach number distribution on the wing is nearly the same in the two cases, however, reaching a peak of approximately 2.05 underneath the vortex. The radial Mach number is shown in Figures 5.49 and 5.50; the cross-flow Mach number is shown in Figures 5.51 and 5.52. The cross-flow Mach number is basically the same in the two cases, but the radial Mach number is very different. In the isentropic case, the radial Mach number reaches a value of 1.8 in the core; in the non-isentropic case it only reaches a value of 1.5. The density, shown in Figures 5.53 and 5.54, reaches a lower value in the core in the non-isentropic case. As with the other variables, the distribution on the wing is the same in the two cases.

In general, the isentropic assumption does not change the flow variables on the wing. It is only in the core of the vortex that the two calculations differ. The radial Mach shows the greatest difference, with a much larger surplus of radial Mach number in the isentropic core. Previously reported isentropic results for the Monnerie and Werlé $\alpha = 10°$ case [48] show that the pitot pressures predicted by the isentropic model are a great deal less realistic than those calculated by the non-isentropic conical Euler model.

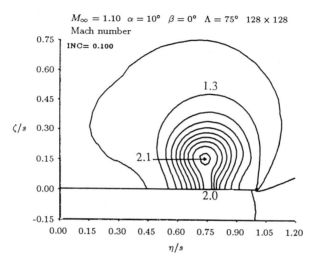

Figure 5.47: Isentropic case — Mach number

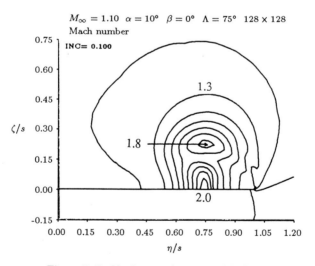

Figure 5.48: Nonisentropic case — Mach number

5.5. LOSSLESS SOLUTIONS TO THE EULER EQUATIONS

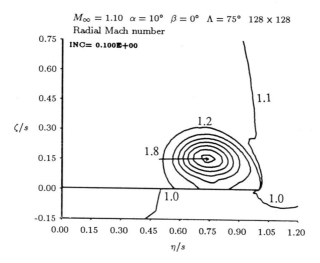

Figure 5.49: Isentropic case — Radial Mach number

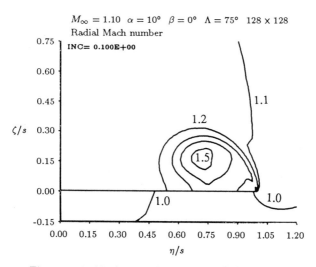

Figure 5.50: Nonisentropic case — Radial Mach number

98 CHAPTER 5. TOTAL PRESSURE LOSSES — A NUMERICAL STUDY

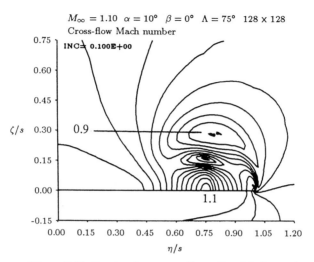

Figure 5.51: Isentropic case — Cross-flow Mach number

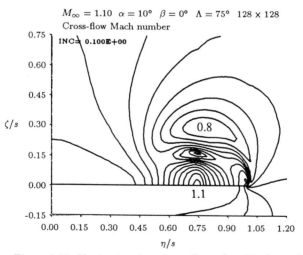

Figure 5.52: Nonisentropic case — Cross-flow Mach number

5.5. LOSSLESS SOLUTIONS TO THE EULER EQUATIONS

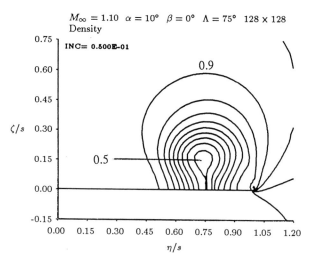

Figure 5.53: Isentropic case — Density

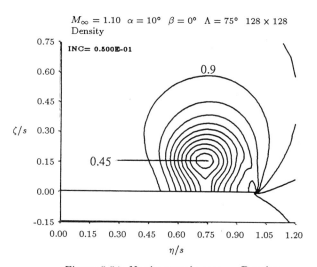

Figure 5.54: Nonisentropic case — Density

5.6 Summary

This chapter has presented a thorough study of the total pressure losses that occur in the calculations. The level of artificial viscosity has been found to have virtually no effect on the level of loss in the core, and very little effect on the macro-scale for the vortex. The micro-scale for the vortex has been found to be extremely sensitive to artificial viscosity level. Grid refinement also affects the micro-scale, with very little effect on the macro-scale or the level of total pressure loss on the core. A fine grid (256×256) result has shown more than one turn of the vortex sheet converging into a core that has approximately the same loss as in the coarser grid cases. The form of the artificial viscosity has been found to have an effect on the core total pressure loss, as have the angle of attack, free-stream Mach number and leading-edge sweep. The losses have been compared to experimentally measured losses, and found to be very realistic. A study of the level of artificial viscosity in the core has been made. It was found that the errors in Crocco's relation are order one in the region of the core on fine and coarse grids, and that the equivalent Reynolds number is on the order of one hundred on the coarsest grid and one thousand on the finer grids. Isentropic results have been presented, calculated using a modified equation set. They have been found to give the same results as the non-isentropic case everywhere except in the core of the vortex. In the core, the non-isentropic results have been shown to be more physically realistic.

Chapter 6

Total Pressure Losses — A Theoretical Model

In this chapter, a model for the total pressure losses in vortical solutions to the Euler equations is presented. It explains the independence of numerical parameters and the dependence upon physical parameters. In the first section, the way that the discrete Euler equations model the feeding sheet will be described. In the second and third sections, the vortex core will be examined. The second section looks at an existing core model developed by Burgers. The third section presents a new core model based on a similarity solution to the Navier-Stokes equations.

6.1 Model for the Feeding Sheet

The loss that occurs in the feeding sheet is explained below in two parts. In the first part, the form that the Euler equations take in the vicinity of the feeding sheet is examined. It is shown that numerical parameters cause the sheet to be spread across several grid points. In the second part, it is shown that a "sheet" that is spread across several points must exhibit a loss. Similar results have been reached independently by Hirschel and Rizzi [21] and by Eberle, Rizzi and Hirschel [13], showing that the "Euler wake" of a lifting wing, obtained by a discrete Euler approach, has the correct kinematic behavior.

To understand the way in which the discretized form of the Euler equations models a vortex sheet, it is first necessary to understand the way in which the partial differential equations themselves model a vortex sheet. The vortex sheet is a weak solution to the

Euler equations [52]. If the Euler equations are written in terms of the Cartesian state and flux vectors,

$$\frac{\partial \mathbf{U}}{\partial t} + \frac{\partial \mathbf{F}}{\partial x} + \frac{\partial \mathbf{G}}{\partial y} + \frac{\partial \mathbf{H}}{\partial z} = 0 \; , \tag{6.1}$$

then a weak solution $\mathbf{U}$ is defined by

$$\iiiint \left[\frac{\partial \mathbf{T}}{\partial t} \cdot \mathbf{U} + \frac{\partial \mathbf{T}}{\partial x} \cdot \mathbf{F} + \frac{\partial \mathbf{T}}{\partial y} \cdot \mathbf{G} + \frac{\partial \mathbf{T}}{\partial z} \cdot \mathbf{H} \right] dx \, dy \, dz \, dt = 0 \tag{6.2}$$

for any vector test function $\mathbf{T}$ that is infinitely differentiable and vanishes at infinity. Carrying out the integration gives

$$U_s \|\mathbf{U}\| + n_x \|\mathbf{F}\| + n_y \|\mathbf{G}\| + n_z \|\mathbf{H}\| = 0 \; , \tag{6.3}$$

in which $\mathbf{n}$ is a vector perpendicular to the surface of discontinuity, U_s is the velocity at which the surface is moving, and the delimiters denote the jump in a quantity across the surface. It should be noted that these relations are invariant to the conical self-similarity assumption, which simply enforces that the surface of discontinuity is generated by rays emanating from the apex of the conical flow. The equations may be rewritten in the form

$$\rho_L [\mathbf{n} \cdot \mathbf{u_L} - U_s] = \rho_R [\mathbf{n} \cdot \mathbf{u_R} - U_s] = M \; , \tag{6.4a}$$

$$M [\mathbf{u_R} - \mathbf{u_L}] = \mathbf{n} [p_L - p_R] \; , \tag{6.4b}$$

$$M [h_{0_R} - h_{0_L}] = 0 \; , \tag{6.4c}$$

in which subscripts L and R denote the left and right sides of the discontinuity, respectively. The momentum equations may be rewritten in a more convenient form by resolving them into normal and tangential coordinates. They become

$$M [\mathbf{n} \cdot \mathbf{u_R} - \mathbf{n} \cdot \mathbf{u_L}] = p_L - p_R \; , \tag{6.5a}$$

$$M [\mathbf{n} \times \mathbf{u_R} - \mathbf{n} \times \mathbf{u_L}] = 0. \tag{6.5b}$$

It is now clear that two types of weak solutions exist:

1. $\mathbf{n} \times \mathbf{u}$ is continuous across the discontinuity. The relations are:

$$\rho_L [\mathbf{n} \cdot \mathbf{u_L} - U_s] = \rho_R [\mathbf{n} \cdot \mathbf{u_R} - U_s] = M \; , \tag{6.6a}$$

$$M [\mathbf{n} \cdot \mathbf{u_R} - \mathbf{n} \cdot \mathbf{u_L}] = p_L - p_R \; , \tag{6.6b}$$

$$h_{0_R} - h_{0_L} = 0 \; , \tag{6.6c}$$

$$\mathbf{n} \times \mathbf{u_R} - \mathbf{n} \times \mathbf{u_L} = 0. \tag{6.6d}$$

6.1. MODEL FOR THE FEEDING SHEET

This describes a shock.

2. $M = 0$. The relations are:

$$\mathbf{n} \cdot \mathbf{u_R} = \mathbf{n} \cdot \mathbf{u_L} = U_s, \tag{6.7a}$$

$$p_R = p_L, \tag{6.7b}$$

with the remaining equations being degenerate. This describes a contact discontinuity, with the tangential velocity, the density or the stagnation enthalpy discontinuous across the sheet. It is the tangential velocity discontinuity that is considered here.

It is the degeneracy in the form of the contact discontinuity weak solution that gives rise to problems in the numerical modeling of a vortex sheet. Unless the sheet is lined up with a grid line, the flux into the cell will spread the sheet, changing it from a discontinuity to a feature with structure [49]. Artificial viscosity will also spread the sheet.

Since the tangential velocity undergoes a change from one side of the "sheet" to the other, its magnitude must reach a minimum somewhere inside the "sheet". This may be shown by looking at a coordinate system which has one direction normal to the sheet and the other two tangential to it. The total pressure is given by

$$p_0 = p \left[1 + \frac{\gamma - 1}{2} M^2 \right]^{\frac{\gamma}{\gamma - 1}}. \tag{6.7c}$$

Using the definition of the stagnation enthalpy, this may be rewritten as

$$p_o = p \left[1 + \frac{u_i u_i}{2h_0 - u_j u_j} \right]^{\frac{\gamma}{\gamma - 1}}. \tag{6.7d}$$

Since the enthalpy, total pressure and static pressure are the same on the two sides of the sheet, the magnitude of the tangential velocity must be the same on the two sides of the sheet. The direction of the velocity vector will differ, however, with the change in direction proportional to the strength of the sheet. Let one coordinate direction be aligned with the mean direction of these velocity vectors. The component of velocity normal to this coordinate must change sign across the sheet, reaching zero somewhere inside the sheet. The total pressure will be a minimum at this point (assuming that the component of velocity parallel to the mean velocity does not appreciably change inside sheet). This minimum in total pressure is set by the strength of the sheet, as measured by the change in tangential velocity across it, which is in turn set by the Kutta condition.

104 CHAPTER 6. TOTAL PRESSURE LOSSES — A THEORETICAL MODEL

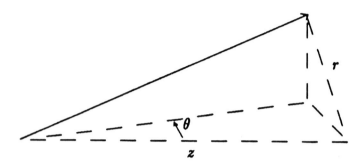

Figure 6.1: Coordinate system for Burgers' vortex

There is therefore a direct relation between the strength of the feeding sheet and the total pressure loss seen within it. For weak sheets, the loss in the sheet is fairly sensitive to computational parameters; for stronger sheets it is much less so.

6.2 Burgers' Vortex

The insensitivity of the level of the loss to mesh spacing and artificial viscosity levels suggests a core model in which the total pressure loss level is independent of Reynolds number. Burgers [9] developed a vortex core model that, it will be shown, has this characteristic. It is an exact solution to the axisymmetric, incompressible Navier-Stokes equations. The coordinate system is a cylindrical (r,θ,z) system, shown in Figure 6.1. The velocity field is given by

$$u = -Ar , \qquad (6.8a)$$

$$v = v(r) , \qquad (6.8b)$$

$$w = 2Az , \qquad (6.8c)$$

where v(r) is to be determined from the θ-momentum equation. The velocities are non-dimensionalized by a reference velocity U, and r and z are non-dimensionalized by a reference length L. This velocity field satisfies the continuity equation,

$$\frac{1}{r}\frac{\partial}{\partial r}(ru) + \frac{\partial w}{\partial z} = 0. \qquad (6.9)$$

6.2. BURGERS' VORTEX

The r-momentum equation (with pressure non-dimensionalized by ρU^2)

$$u\frac{\partial u}{\partial r} + w\frac{\partial u}{\partial z} - \frac{v^2}{r} = -\frac{\partial p}{\partial r} + \frac{1}{Re}\left(\frac{\partial^2 u}{\partial r^2} + \frac{1}{r}\frac{\partial u}{\partial r} - \frac{u}{r^2} + \frac{\partial^2 u}{\partial z^2}\right) +$$
$$+ \frac{1}{3Re}\frac{\partial}{\partial r}\left(\frac{\partial u}{\partial r} + \frac{u}{r} + \frac{\partial w}{\partial z}\right) \quad (6.10)$$

reduces, for the above choices of u, v and w to

$$\frac{\partial p}{\partial r} = \frac{v^2}{r} - u\frac{\partial u}{\partial r}. \quad (6.11)$$

The z-momentum equation

$$u\frac{\partial w}{\partial r} + w\frac{\partial w}{\partial z} = -\frac{\partial p}{\partial z} + \frac{1}{Re}\left(\frac{\partial^2 w}{\partial r^2} + \frac{1}{r}\frac{\partial w}{\partial r} + \frac{\partial^2 w}{\partial z^2}\right) +$$
$$+ \frac{1}{3Re}\frac{\partial}{\partial z}\left(\frac{\partial u}{\partial r} + \frac{u}{r} + \frac{\partial w}{\partial z}\right) \quad (6.12)$$

reduces to

$$\frac{\partial p}{\partial z} = -w\frac{\partial w}{\partial z}. \quad (6.13)$$

The θ-momentum equation

$$w\frac{\partial v}{\partial z} + u\frac{\partial v}{\partial r} + \frac{uv}{r} = \frac{1}{Re}\left(\frac{\partial^2 v}{\partial r^2} + \frac{1}{r}\frac{\partial v}{\partial r} - \frac{v}{r^2} + \frac{\partial^2 v}{\partial z^2}\right) \quad (6.14)$$

becomes

$$u\frac{\partial v}{\partial r} + \frac{uv}{r} = \frac{1}{Re}\left(\frac{\partial^2 v}{\partial r^2} + \frac{1}{r}\frac{\partial v}{\partial r} - \frac{v}{r^2}\right). \quad (6.15)$$

Burgers' original analysis was carried out in unscaled variables. The relation between the total pressure and the Reynolds number may be seen most easily, however, by rescaling the equations. Introducing scaled variables

$$\hat{u} = u\sqrt{\frac{ARe}{2}}, \quad (6.16a)$$

$$\hat{v} = \frac{Av}{2}, \quad (6.16b)$$

$$\hat{w} = w, \quad (6.16c)$$

$$\hat{p} = p, \quad (6.16d)$$

$$\hat{r} = r\sqrt{\frac{ARe}{2}}, \quad (6.16e)$$

$$\hat{z} = z, \quad (6.16f)$$

leaves the continuity equation unchanged. The θ-momentum equation becomes, with the assumed form for u,

$$\frac{\partial^2 \hat{v}}{\partial \hat{r}^2} + \left(\frac{1}{\hat{r}} + 2\hat{r}\right)\frac{\partial \hat{v}}{\partial \hat{r}} + \left(2 - \frac{1}{\hat{r}^2}\right)\hat{v} = 0. \qquad (6.17)$$

Applying the boundary conditions $\hat{v}(0) = 0$ and $\hat{r}\hat{v} \to A\hat{\Gamma}/2$ as $\hat{r} \to \infty$ gives

$$\hat{v} = \frac{A\hat{\Gamma}}{2\hat{r}}\left(1 - e^{-\hat{r}^2}\right), \qquad (6.18)$$

where $\hat{\Gamma}$ is a scaled circulation, related to the physical circulation Γ by

$$\hat{\Gamma} = \frac{\Gamma}{2\pi}\sqrt{\frac{ARe}{2}}. \qquad (6.19)$$

The r and z momentum equations, when integrated, yield the relation for the static pressure

$$\hat{p} = \frac{4}{A^2}\int \frac{\hat{v}^2}{\hat{r}}d\hat{r} - \frac{1}{Re}A\hat{r}^2 - 2A^2\hat{z}^2. \qquad (6.20)$$

From this the total pressure may be calculated, giving

$$p_0 = \frac{4}{A^2}\left[\int \frac{\hat{v}^2}{\hat{r}}d\hat{r} + \frac{\hat{v}^2}{2}\right] = \qquad (6.21\text{a})$$

$$= \hat{\Gamma}^2\left[\int \left(\frac{1-e^{-\hat{r}^2}}{\hat{r}}\right)^2\frac{d\hat{r}}{\hat{r}} + \frac{1}{2}\left(\frac{1-e^{-\hat{r}^2}}{\hat{r}}\right)^2\right]. \qquad (6.21\text{b})$$

Figure 6.2 is a plot of $p_0/\hat{\Gamma}^2$ as a function of $\hat{r}$ which holds for any Reynolds number. This means that the level of the total pressure loss in the core is set solely by $\hat{\Gamma}$, where $\hat{\Gamma} \sim \Gamma\sqrt{ARe}$. The position of the "edge" of the vortex, r_e, scales with $1/\sqrt{ARe}$, which implies that $\Gamma \sim v_e r_e \sim v_e/\sqrt{ARe}$. Thus, $\hat{\Gamma} \sim v_e$, and for a given edge swirl velocity, the total pressure loss level in a Burgers' vortex is independent of Reynolds number. If the value of the edge velocity, v_e is known, the *magnitude* of the total pressure loss in the core is determined. This is not an obvious result — the static pressure depends on the Reynolds number, as does the radial velocity. The *distribution* of the total pressure loss does depend on the Reynolds number; it scales with $1/\sqrt{Re}$.

6.3 Model for Total Pressure Loss

While Burgers' vortex exhibits the desired independence of Reynolds number for the level of the total pressure loss, the velocity field is not a good model for the leading-edge vortex flows presented here. This section derives a conical analog to Burgers' vortex that

6.3. MODEL FOR TOTAL PRESSURE LOSS

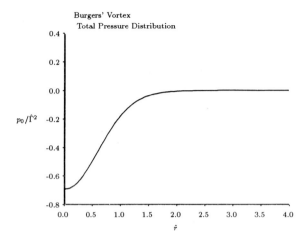

Figure 6.2: Total pressure distribution in Burgers' vortex — $p_0/\hat{\Gamma}^2$ vs $\hat{r}$

exhibits the same behavior for the total pressure. The assumption is made that, in the circumferential momentum equation, the viscous terms are not negligible but of the same order as the convective terms. The vortex core is also assumed to be slender. The flow is assumed to be conical. While a viscous flow may not be truly conical, this is a good model for the numerical scheme presented in Chapter 3, since the added artificial viscosity is conical.

The equations of motion for axisymmetric, viscous, steady, incompressible flow are, in a cylindrical coordinate system, the continuity equation,

$$\frac{1}{r}\frac{\partial}{\partial r}(ru) + \frac{\partial w}{\partial z} = 0, \tag{6.22}$$

the r momentum equation,

$$u\frac{\partial u}{\partial r} + w\frac{\partial u}{\partial z} - \frac{v^2}{r} = -\frac{\partial p}{\partial r} + \frac{1}{Re}\left[\frac{1}{r}\frac{\partial}{\partial r}\left(r\frac{\partial u}{\partial r}\right) + \frac{\partial^2 u}{\partial z^2} - \frac{u}{r^2}\right], \tag{6.23}$$

the θ momentum equation,

$$u\frac{\partial v}{\partial r} + w\frac{\partial v}{\partial z} + \frac{uv}{r} = \frac{1}{Re}\left[\frac{1}{r}\frac{\partial}{\partial r}\left(r\frac{\partial v}{\partial r}\right) + \frac{\partial^2 v}{\partial z^2} - \frac{v}{r^2}\right], \tag{6.24}$$

and the z momentum equation

$$u\frac{\partial w}{\partial r} + w\frac{\partial w}{\partial z} = -\frac{\partial p}{\partial z} + \frac{1}{Re}\left[\frac{1}{r}\frac{\partial}{\partial r}\left(r\frac{\partial w}{\partial r}\right) + \frac{\partial^2 w}{\partial z^2}\right], \tag{6.25}$$

where the Reynolds number is based on a reference length L and a reference velocity W. Introducing the conical variable $\phi = r/z$ and assuming conical self-similarity, the continuity equation becomes

$$-\phi\left(w - \frac{u}{\phi}\right)' + 2\frac{u}{\phi} = 0 , \qquad (6.26)$$

the r momentum equation becomes

$$-\phi\left(w - \frac{u}{\phi}\right)u' - \frac{v^2}{\phi} = -p' + \frac{1}{Re_z}\left[\left(1 + \phi^2\right)u'' + \left(\frac{1}{\phi} + 2\phi\right)u' - \frac{u}{\phi^2}\right] , \qquad (6.27)$$

the θ momentum equation becomes

$$-\phi\left(w - \frac{u}{\phi}\right)v' + \frac{uv}{\phi} = \frac{1}{Re_z}\left[\left(1 + \phi^2\right)v'' + \left(\frac{1}{\phi} + 2\phi\right)v' - \frac{v}{\phi^2}\right] , \qquad (6.28)$$

and the z momentum equation becomes

$$-\phi\left(w - \frac{u}{\phi}\right)w' = \phi p' + \frac{1}{Re_z}\left[\left(1 + \phi^2\right)w'' + \left(\frac{1}{\phi} + 2\phi\right)w'\right] , \qquad (6.29)$$

where Re_z is a Reynolds number based on the local value of z and the primes denote differentiation by ϕ. Since the Reynolds number has a z dependence, the conical assumption does not truly hold. Conical self-similarity may therefore hold only in a "local" sense.

The r and z momentum equations may be rewritten as an axial momentum equation, in which there are no pressure terms, and a ϕ momentum equation, in which there are no viscous terms. The three momentum equations may then be written as a radial momentum equation:

$$-\frac{\phi^2}{4}\left[\left(w - \frac{u}{\phi}\right)^2\right]' + \frac{v^2}{\phi} = \left(1 + \phi^2\right)p' , \qquad (6.30)$$

a circumferential momentum equation:

$$\left(1 + \phi^2\right)v'' + \left[\frac{1}{\phi} + 2\phi + \phi\left(w - \frac{u}{\phi}\right)Re_z\right]v' - \left[\frac{1}{\phi^2} + \frac{\phi}{2}\left(w - \frac{u}{\phi}\right)'Re_z\right]v = 0 \qquad (6.31)$$

and an axial momentum equation:

$$\left(1 + \phi^2\right)(w + \phi u)'' + \left[\frac{1}{\phi} + 2\phi + \phi\left(w - \frac{u}{\phi}\right)Re_z\right](w + \phi u)' -$$

$$- \left(\phi^2 + \phi^4\right)\left(w - \frac{u}{\phi}\right)'' - 3\left(\phi + \phi^3\right)\left(w - \frac{u}{\phi}\right)' + Re_z\left[v^2 - \phi\left(w - \frac{u}{\phi}\right)u\right] = 0 . \qquad (6.32)$$

There are two limiting cases to these equations which are of interest. Both are high Reynolds number, slender core limits based on Re_z and Φ, a reference value for ϕ. The

6.3. MODEL FOR TOTAL PRESSURE LOSS

first limit is the inviscid limit, first examined by Hall [19], who looked at the inviscid equations of motion. In it, $Re_z \to \infty$ and $\Phi \to 0$ such that $Re_z \Phi^2 \to \infty$. In this limit, the continuity equation is unchanged:

$$-\phi \left(w - \frac{u}{\phi}\right)' + 2\frac{u}{\phi} = 0, \tag{6.33}$$

the radial momentum equation reduces to a balance of pressure and centrifugal force

$$p' = \frac{v^2}{\phi}, \tag{6.34}$$

the circumferential momentum equation becomes

$$\left(w - \frac{u}{\phi}\right) v' = \frac{1}{2}\left(w - \frac{u}{\phi}\right)' v, \tag{6.35}$$

and the axial momentum equation reduces to

$$\phi \left(w - \frac{u}{\phi}\right) w' + v^2 = 0. \tag{6.36}$$

Applying the boundary conditions

$$u = 0 \quad at \quad \phi = 0,$$

$$v = V \quad at \quad \phi = \Phi,$$

$$w = 1 \quad at \quad \phi = \Phi$$

allows the equations to be integrated exactly, giving

$$u = \frac{\phi}{2\Phi}\left(1 - \sqrt{1 + 2V^2}\right), \tag{6.37a}$$

$$v = \sqrt{V^2 - \left(1 - \sqrt{1 + 2V^2}\right)^2 \log \frac{\phi}{\Phi}}, \tag{6.37b}$$

$$w = 1 + \left(1 - \sqrt{1 + 2V^2}\right) \log \frac{\phi}{\Phi}. \tag{6.37c}$$

Integrating the radial momentum equation gives

$$C_p = 2V^2 \log \frac{\phi}{\Phi} - \left(1 - \sqrt{1 + 2V^2}\right)^2 \log^2 \frac{\phi}{\Phi}. \tag{6.38}$$

The total pressure relation consistent with the limit taken is

$$p_0 = p + \frac{1}{2}\left(v^2 + w^2\right), \tag{6.39}$$

CHAPTER 6. TOTAL PRESSURE LOSSES — A THEORETICAL MODEL

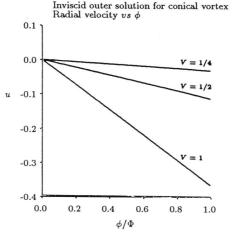

Figure 6.3: Outer solution – u vs ϕ

which is constant for the given pressure and velocity fields. This solution is plotted, for three values of V, in Figures 6.3-6.6. Luckring [31] extended Hall's analysis, keeping the higher-order terms in Φ, and found only small changes due to the slenderness assumption.

As can be seen, the inviscid solution has a log-like singularity at the origin. Hall [19] and Stewartson and Hall [61] removed this singularity by matching a viscous, non-conical inner solution to this inviscid conical outer solution. A different approach is taken here.

The approach taken here is based on the viscous equations. A different limit of Equation 6.26 and Equations 6.30-6.32 is taken, one in which the viscous terms and convective terms in the θ momentum equation are balanced. Thus, $Re_z \to \infty$ and $\Phi \to 0$ such that $Re_z \Phi^2 \to 1$. Introducing the scaled variables

$$\hat{u} = u\sqrt{Re_z}, \tag{6.38a}$$

$$\hat{v} = v, \tag{6.38b}$$

$$\hat{w} = w, \tag{6.38c}$$

$$\hat{p} = p, \tag{6.38d}$$

$$\hat{\phi} = \phi\sqrt{Re_z}, \tag{6.38e}$$

and dropping the high-order terms, the continuity equation becomes

$$-\hat{\phi}\left(\hat{w} - \frac{\hat{u}}{\hat{\phi}}\right)' + 2\frac{\hat{u}}{\hat{\phi}} = 0, \tag{6.39}$$

6.3. MODEL FOR TOTAL PRESSURE LOSS

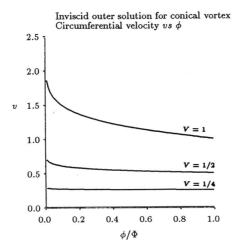

Figure 6.4: Outer solution – v vs ϕ

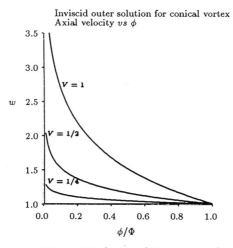

Figure 6.5: Outer solution – w vs ϕ

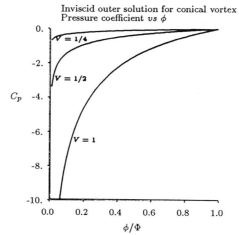

Figure 6.6: Outer solution – C_p vs ϕ

the radial momentum equation becomes

$$\hat{p}' = \frac{\hat{v}^2}{\hat{\phi}}, \tag{6.40}$$

the circumferential momentum equation becomes

$$\hat{v}'' + \left[\frac{1}{\hat{\phi}} + \hat{\phi}\left(\hat{w} - \frac{\hat{u}}{\hat{\phi}}\right)\right]\hat{v}' - \left[\frac{1}{\hat{\phi}^2} + \frac{\hat{\phi}}{2}\left(\hat{w} - \frac{\hat{u}}{\hat{\phi}}\right)'\right]\hat{v} = 0, \tag{6.41}$$

and the axial momentum equation becomes

$$\hat{w}'' + \left[\frac{1}{\hat{\phi}} + \hat{\phi}\left(\hat{w} - \frac{\hat{u}}{\hat{\phi}}\right)\right]\hat{w}' + \hat{v}^2 = 0, \tag{6.42}$$

where the primes now denote differentiation with respect to $\hat{\phi}$.

As written, there are no parameters in these equations — they enter only through the boundary conditions. The equations require one boundary condition each for p and u, and two each for v and w. The boundary conditions chosen for the velocities are

$$\hat{u} = 0 \quad at \quad \hat{\phi} = 0 \;,$$

$$\hat{v} = 0 \quad at \quad \hat{\phi} = 0 \;,$$

$$\hat{w}' = 0 \quad at \quad \hat{\phi} = 0 \;,$$

$$\hat{w} = 1 \quad at \quad \hat{\phi} = \hat{\phi}_{max} \;,$$

$$\hat{\phi}\hat{v} \sim \hat{\Gamma} \quad at \quad \hat{\phi} = \hat{\phi}_{max} \longrightarrow \hat{v}' = -\frac{\hat{\Gamma}}{\hat{\phi}^2} \quad at \quad \hat{\phi} = \hat{\phi}_{max}.$$

6.3. MODEL FOR TOTAL PRESSURE LOSS

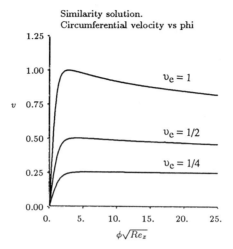

Figure 6.7: Conical vortex similarity solution $\hat{v}$ vs $\hat{\phi}$

This system of equations for the velocities is rewritten as a set of five first order equations, linearized and solved using a box scheme [28]. Solutions are presented in Figures 6.7-6.11. The circumferential and axial velocities are plotted, along with the vorticity and the coefficients of static and total pressure. The vorticity is normalized by $z/\sqrt{Re_z}$. There are two parameters in the boundary conditions, $\hat{\Gamma}$ and $\hat{\phi}_{max}$. Solutions may be characterized by a single parameter, however. This parameter is v_e, the maximum swirl velocity of the vortex. The scaled circulation $\hat{\Gamma}$ is related to the physical circulation Γ by the relation $\hat{\Gamma} \sim 2\pi\Gamma\sqrt{Re_z}$. Thus, as was the case with Burgers' vortex, it is the swirl velocity v_e that sets the level of total pressure loss in the core. That this is true may be seen in Table 6.3, where the maximum circumferential velocity v_e and the minimum total pressure $C_{h_{min}}$ are presented for the three cases. A plot of $C_{h_{min}}$ vs v_e is shown in Figure 6.12. The solid line is the result of the conical model; the symbols are the curve $1 - 2v_e^2$. As with Burgers' vortex, the total pressure loss is proportional to the square of the edge swirl velocity of the vortex.

This model fits very well with what is seen in the conical Euler solutions presented in this book. The implications of the model are:

1. Total pressure loss level is independent of level of viscosity;

2. Total pressure loss level is directly related to swirl velocity;

114 CHAPTER 6. TOTAL PRESSURE LOSSES — A THEORETICAL MODEL

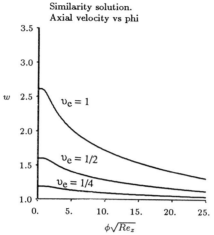

Figure 6.8: Conical vortex similarity solution $\hat{w}$ vs $\hat{\phi}$

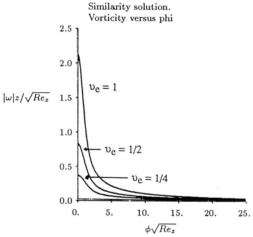

Figure 6.9: Conical vortex similarity solution $\hat{\omega}$ vs $\hat{\phi}$

Table 6.1: $C_{h_{min}}$ and v_e as a function of $\hat{\Gamma}$ and $\hat{\phi}_{max}$

$\hat{\phi}_{max}$	$\hat{\Gamma}$	v_e	$C_{h_{min}}$
25	1.00	0.514	0.467
50	1.00	0.392	0.696
50	1.85	0.515	0.468

6.3. MODEL FOR TOTAL PRESSURE LOSS

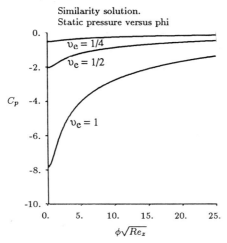

Figure 6.10: Conical vortex similarity solution C_p vs $\hat{\phi}$

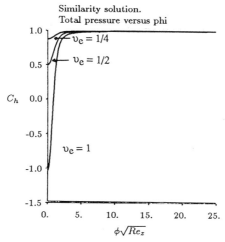

Figure 6.11: Conical vortex similarity solution C_h vs $\hat{\phi}$

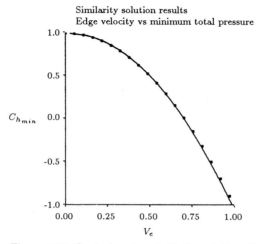

Figure 6.12: Conical vortex similarity solution $C_{h_{min}}$ vs v_e

3. Vortex size scales with the square root of the Reynolds number.

In Chapter 5, it was shown that:

1. The total pressure loss level is independent of the level of artificial viscosity and grid refinement;

2. The maximum flow angularity above the vortex, a measure of the swirl velocity, is constant with grid refinement;

3. There are two scales for the size of the vortex — the larger scale is basically independent of the Reynolds number (or the numerical parameters that set the level of the diffusive terms), the smaller scale is dependent upon the Reynolds number.

6.4 Summary

This chapter has presented models for the total pressure loss in the feeding sheet and the vortex core. The model for the feeding sheet is based on the discretized form of the contact discontinuity weak solution to the Euler equations. The loss in the sheet has been seen to be brought about by the spreading of the sheet due to grid and artificial viscosity effects. The level of the loss has been seen to be set by the strength of the sheet, due to the

6.4. SUMMARY

deficit in velocity it. The vortex core model has been introduced by showing that Burgers' vortex, an exact solution to the axisymmetric, incompressible Navier-Stokes equations, exhibits a core total pressure loss whose level is set by the edge velocity of the vortex, and is independent of Reynolds number. A new core model, based on a high Reynolds number, slender core limit similarity solution to the axisymmetric, incompressible, conical Navier-Stokes equations has been presented. The new core model exhibits a core total pressure loss whose level is independent of Reynolds number, and whose distribution scales with the square root of the Reynolds number. This has been shown to be consistent with the conical Euler results presented in Chapter 5.

Chapter 7

Comparison of Computations and Experiments

This chapter presents a comparison of solutions of the conical Euler equations and the results of a series of experiments carried out at NASA Langley Research Center by Miller and Wood [36,50] The first section describes the models and the set-up used in the experiments. Three symmetric flat plate cases, with various values of free-stream Mach number, angle of attack and leading-edge sweep, are presented in the second section. Two vortex flap cases are presented in the second section. Two asymmetric flat plate cases are presented in the final section.

7.1 Experimental Setup and Models

The experimental results presented in this chapter come from experiments carried out at NASA Langley Research Center. A full description of the models, the tunnel and the data acquisition techniques is given by Miller and Wood [36]. A summary of the models and test conditions is given here.

The flat plate models consisted of a series of four delta wings of different sweeps, each of which had a span of 12 inches. The vortex flap models consisted of a series of four delta wings, three with camber and one without, each with a leading-edge sweep of 75° and the same upper surface wetted area. The reference flat delta wing had a span of 18 inches and a length of 33.588 inches. Wing camber was generated for the other three wings by a deflection of the outboard 30% of the local wing semi-span to angles of 5°, 10°

and 15° measured in the streamwise direction. To minimize the effects of airfoil shape and thickness, the leading-edges of all models were made sharp and the upper surfaces were flat. A minimum balance housing was incorporated into the models.

The leeward surface of each model was instrumented with rows of 19 evenly-spaced pressure orifices. For the vortex flap models, the rows were located at 10%, 20%, 30%, 60%, 80% and 90% of the model length. For the flat plate models, only one row of orifices was used, located 1 inch forward of the trailing-edge. Pressure orifices were distributed over the semi-span of all wings at equal distances measured along the model upper surface from 0% to 90% of the local semi-span. Pressure data were obtained from a scanning-valve pressure transducer mounted external to the wind tunnel.

The models were connected to the permanent model-actuating system of the tunnel by a six-component strain gauge balance and sting arrangement. The balance was housed in a minimum body which was symmetrically integrated into the wing geometry. During the test, the angle of attack was measured with an accelerometer located in the permanent model-actuating system and was corrected for tunnel flow angularity. Sting deflection effects were accounted for during the test in the flat plate cases. Force and moment data were corrected to free-stream static pressure at the balance chamber.

The tests were conducted in the low Mach number test section of the Langley Unitary Plan Wind Tunnel, which is a variable Mach number, variable pressure, continuous-flow, supersonic tunnel. The test section is approximately 4 feet × 4 feet × 7 feet long. This facility is described in more detail by Jackson *et al* [23]. The nominal Reynolds number was 2×10^6 per foot. To ensure fully turbulent boundary-layer flow over the model at attached flow conditions, according to the guidelines set forth by Braslow and Knox [5], transition strips composed of # 60 carborundum grit were sprinkled on the upper surface 0.2 inches behind the model leading edge (measured normal to the leading edge). The transition strips were approximately 0.0625 inches wide.

7.2 Flat plate cases

This section presents the results for a flat plate at zero yaw. The cases presented are:

1. $M_\infty = 1.7$, $\alpha = 12°$, $\Lambda = 75°$, 128 × 128 equivalent refinement;

2. $M_\infty = 2.0$, $\alpha = 20°$, $\Lambda = 60°$, 128 × 128 equivalent refinement;

7.2. FLAT PLATE CASES

3. $M_\infty = 2.8$, $\alpha = 20°$, $\Lambda = 75°$, 128×128 equivalent refinement.

The first case results in an elliptical vortex with a cross-flow shock underneath it. The second results in a long, thin vortex that lies directly on the wing and a cross-flow shock above the vortex. The third results in a complex shock-vortex system with three cross-flow shocks.

7.2.1 Case 1 — $M_\infty = 1.7$, $\alpha = 12°$, $\Lambda = 75°$

This case was run on an equivalent 128×128 grid, shown in Figure 7.1. The free-stream Mach number, angle of attack and leading-edge sweep correspond to a normal Mach number $M_N = 0.44$ and a normal angle of attack $\alpha_N = 39.4°$. The cross-flow streamlines (Figure 7.2) show the vortex, the node at the windward symmetry point, and saddle points at the leeward symmetry point and on the windward side of the wing near the leading-edge. The underside of the vortex shows a reverse curvature, which is indicative of a cross-flow shock underneath the vortex. The flow angularity (Figure 7.3) reaches 45° under the vortex, and 25° above it. The high gradient in angularity under the vortex is evidence of the reverse cross-flow shock there. The shock is clearly seen in the cross-flow velocity vectors (Figures 7.4 and 7.5). The tuft patterns from the computation (Figure 7.6) and the experiment (Figure 7.7) show the large outboard component of the flow under the vortex.

The vapor screen for this case (Figure 7.8) shows the primary vortex, feeding sheet, cross-flow shock and secondary vortex. Each shows up as a dark region in the vapor screen, having less water condensate than other regions of the flow. The reason for the lack of condensate in these regions has not been well explained to date [36]. The contours of total pressure loss (Figure 7.10) show the vortex, feeding sheet and cross-flow shock; the viscous-induced secondary vortex is not modeled. On the wing, the total pressure loss is zero except on the leeward side, outboard of the cross-flow shock (See Figure 7.9).

The pressure coefficient contours (Figure 7.11) show the low pressure region in the core, the expansion at the leading-edge and the cross-flow shock. There is a large constant pressure region outboard of the cross-flow shock. The computed pressure on the wing (solid line in Figure 7.12) compares well with the measured pressure (symbols in Figure 7.12) inboard and outboard of the vortex, but the suction peak under the pri-

122   CHAPTER 7. COMPARISON OF COMPUTATIONS AND EXPERIMENTS

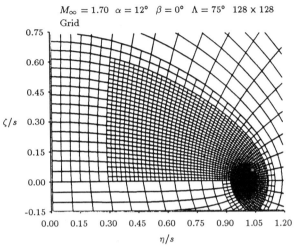

Figure 7.1: Grid — $M_\infty = 1.7$, $\alpha = 12°$, $\Lambda = 75°$

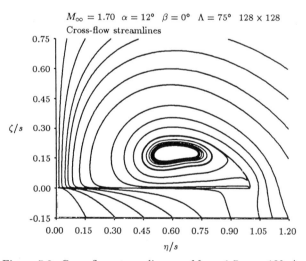

Figure 7.2: Cross-flow streamlines — $M_\infty = 1.7$, $\alpha = 12°$, $\Lambda = 75°$

7.2. FLAT PLATE CASES

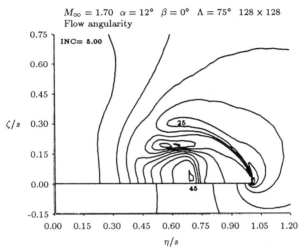

Figure 7.3: Flow angularity — $M_\infty = 1.7$, $\alpha = 12°$, $\Lambda = 75°$

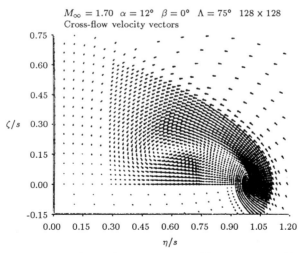

Figure 7.4: Cross-flow velocity vectors — $M_\infty = 1.7$, $\alpha = 12°$, $\Lambda = 75°$

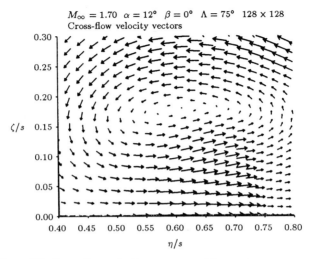

Figure 7.5: Cross-flow velocity vectors — $M_\infty = 1.7$, $\alpha = 12°$, $\Lambda = 75°$

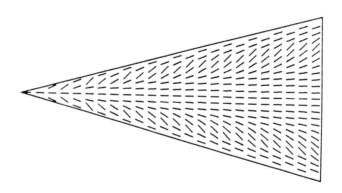

Figure 7.6: Computed tuft patterns — $M_\infty = 1.7$, $\alpha = 12°$, $\Lambda = 75°$

7.2. FLAT PLATE CASES

Figure 7.7: Experimental tuft patterns — $M_\infty = 1.7$, $\alpha = 12°$, $\Lambda = 75°$

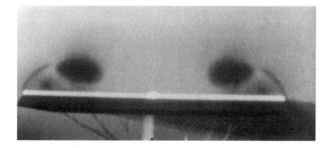

Figure 7.8: Experimental vapor screen — $M_\infty = 1.7$, $\alpha = 12°$, $\Lambda = 75°$

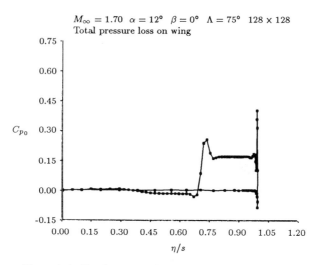

Figure 7.9: Total pressure loss — $M_\infty = 1.7$, $\alpha = 12°$, $\Lambda = 75°$

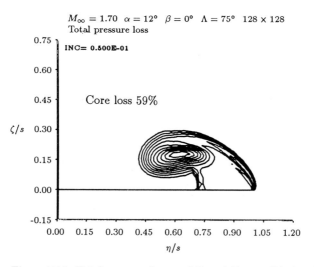

Figure 7.10: Total pressure loss — $M_\infty = 1.7$, $\alpha = 12°$, $\Lambda = 75°$

7.2. FLAT PLATE CASES

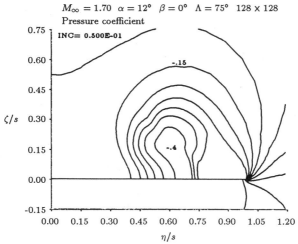

Figure 7.11: Pressure coefficient — $M_\infty = 1.7$, $\alpha = 12°$, $\Lambda = 75°$

mary vortex is over-predicted. Also, the measured pressure shows a weak expansion and recompression under the secondary vortex that is not modeled by the computations.

The cross-flow Mach number on the windward side of the wing (Figure 7.13) is zero at the symmetry point, and remains low to the cross-flow stagnation point just inboard of the leading-edge. On the leeward side, it is zero at the symmetry point, expanding rapidly under the vortex, recompressing through the cross-flow shock, and decreasing outboard to the leading-edge. The cross-flow Mach number reaches 1.4 above the vortex (See Figure 7.14) and 1.7 below it, just inboard of the cross-flow shock. The high gradients in the sheet and the core are shown. The gradients in radial Mach number are much lower (Figure 7.15). The total Mach number (Figure 7.16) reaches a value of 3.2 above the core, and a value of 4.0 just upstream of the cross-flow shock.

7.2.2 Case 2 — $M_\infty = 2.0$, $\alpha = 20°$, $\Lambda = 60°$

This case, which corresponds to a normal Mach number $M_N = 1.01$ and a normal angle of attack $\alpha = 36.1°$, was run on a grid of 128×128 equivalent global refinement, shown in Figure 7.17. The unique topology of this flow may be seen in the cross-flow velocity vectors, Figures 7.18-7.20. There is a long, thin vortex that resides directly on the wing, with a cross-flow shock sitting on top of the vortex. The cross-flow shock is highly curved, leading to two very weak counter-rotating vortices inboard of the kink in the shock. The

128 CHAPTER 7. COMPARISON OF COMPUTATIONS AND EXPERIMENTS

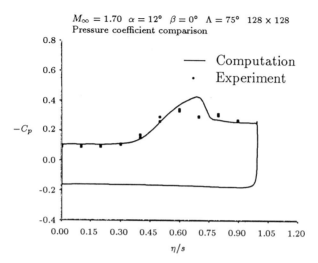

Figure 7.12: Pressure coefficient comparison — $M_\infty = 1.7$, $\alpha = 12°$, $\Lambda = 75°$

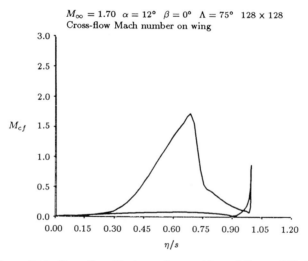

Figure 7.13: Cross-flow Mach number — $M_\infty = 1.7$, $\alpha = 12°$, $\Lambda = 75°$

7.2. FLAT PLATE CASES

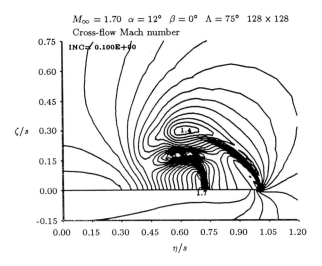

Figure 7.14: Cross-flow Mach number — $M_\infty = 1.7$, $\alpha = 12°$, $\Lambda = 75°$

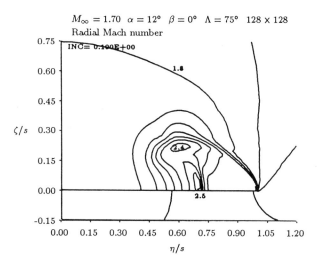

Figure 7.15: Radial Mach number — $M_\infty = 1.7$, $\alpha = 12°$, $\Lambda = 75°$

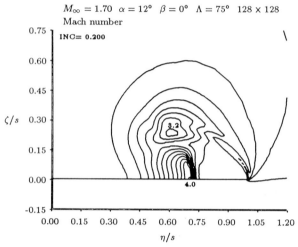

Figure 7.16: Mach number — $M_\infty = 1.7$, $\alpha = 12°$, $\Lambda = 75°$

flow angularity contours, Figure 7.21, show the high degree of turning under the vortex, and the gradient in angularity at the cross-flow shock. The tuft patterns (Figures 7.22 and 7.23) show the extent of the vortex.

The vapor screen shows the vortex on the wing and evidence of the cross-flow shock above the vortex. Both of these show up in the total pressure loss contours shown in Figure 7.26. The cross-flow shock has maximum total pressure loss of 35% downstream of it. Losses in the vortex are as high as 90%. The pressure gradients are low in this case (Figure 7.27), with the expansion at the leading-edge showing up as the most distinct feature in the flow. The pressure coefficient agrees well with experiment (Figure 7.28), although the suction under the vortex is over-predicted near the leading-edge. The cross-flow Mach number (Figures 7.29 and 7.30) is low and nearly constant on the windward side of the wing. On the leeward side, the cross-flow under the vortex is supersonic, with a slight suction peak on the outboard portion of the wing. The flow inboard of the vortex, which has passed over the vortex and through the cross-flow shock above the vortex, undergoes an expansion and recompression on its way to the leeward symmetry point. The cross-flow reaches a Mach number of 2.6 above the vortex.

7.2. FLAT PLATE CASES

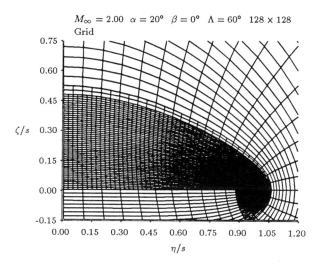

Figure 7.17: Grid — $M_\infty = 2.0$, $\alpha = 20°$, $\Lambda = 60°$

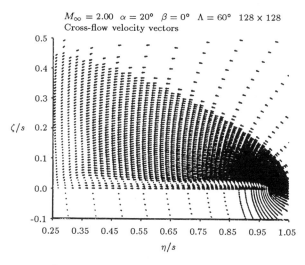

Figure 7.18: Cross-flow velocity vectors — $M_\infty = 2.0$, $\alpha = 20°$, $\Lambda = 60°$

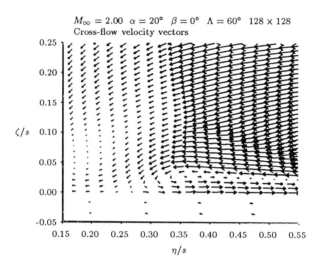

Figure 7.19: Cross-flow velocity vectors — $M_\infty = 2.0$, $\alpha = 20°$, $\Lambda = 60°$

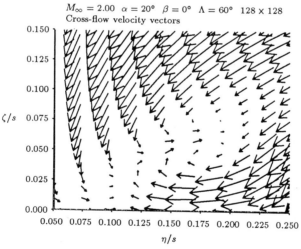

Figure 7.20: Cross-flow velocity vectors — $M_\infty = 2.0$, $\alpha = 20°$, $\Lambda = 60°$

7.2. FLAT PLATE CASES

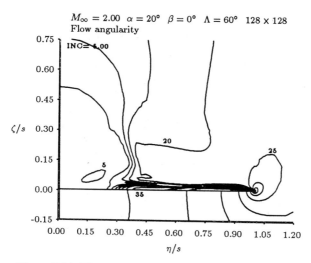

Figure 7.21: Flow angularity — $M_\infty = 2.0$, $\alpha = 20°$, $\Lambda = 60°$

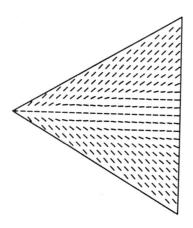

Figure 7.22: Computed tuft patterns — $M_\infty = 2.0$, $\alpha = 20°$, $\Lambda = 60°$

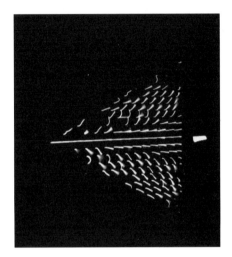

Figure 7.23: Experimental tuft patterns — $M_\infty = 2.0$, $\alpha = 20°$, $\Lambda = 60°$

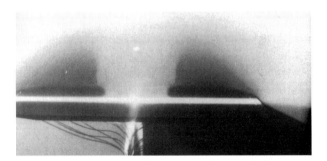

Figure 7.24: Experimental vapor screen — $M_\infty = 2.0$, $\alpha = 20°$, $\Lambda = 60°$

7.2. FLAT PLATE CASES

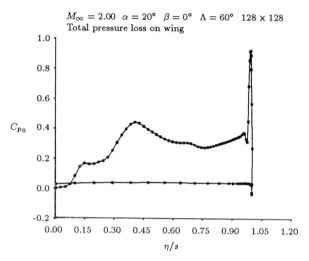

Figure 7.25: Total pressure loss — $M_\infty = 2.0$, $\alpha = 20°$, $\Lambda = 60°$

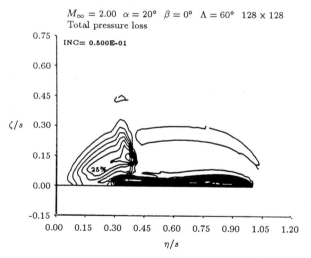

Figure 7.26: Total pressure loss — $M_\infty = 2.0$, $\alpha = 20°$, $\Lambda = 60°$

136   CHAPTER 7. COMPARISON OF COMPUTATIONS AND EXPERIMENTS

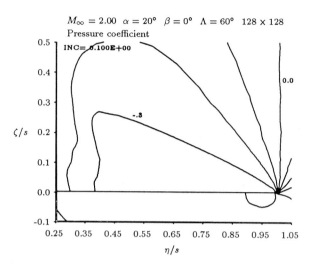

Figure 7.27: Pressure coefficient — $M_\infty = 2.0$, $\alpha = 20°$, $\Lambda = 60°$

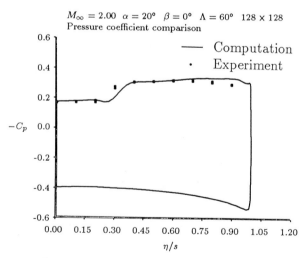

Figure 7.28: Pressure coefficient comparison — $M_\infty = 2.0$, $\alpha = 20°$, $\Lambda = 60°$

7.2. FLAT PLATE CASES

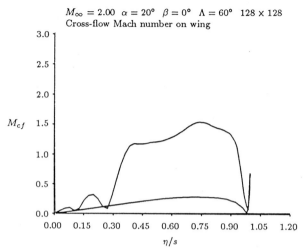

Figure 7.29: Cross-flow Mach number — $M_\infty = 2.0$, $\alpha = 20°$, $\Lambda = 60°$

7.2.3 Case 3 — $M = 2.8$, $\alpha = 20°$, $\Lambda = 75°$

This case was run on an equivalent 128 × 128 grid, shown in Figure 7.32. The cross-flow streamlines (Figure 7.33) show a node at the windward symmetry point and saddle points at the leeward symmetry point, the leading-edge and on the outboard portion of the windward side of the wing. The vortex has a high degree of reverse curvature, suggesting a strong cross-flow shock underneath the vortex. The inflow velocities in the vortex are extremely small; the streamline integration reached a limit cycle even when very small steps (Δs of order 10^{-3}) were taken. There are high gradients in the angularity (Figure 7.34) in the feeding sheet and in the cross-flow shocks above and below the vortex. The sheet and three shocks — above, below and inboard the vortex — are clear in the cross-flow velocity plots (Figures 7.35 and 7.36). The computed and experimental tufts (Figures 7.37 and 7.38) show the high angularity under the vortex, and the way in which the cross-flow shock turns the flow to be parallel to the leading-edge.

The vapor screen for this case (Figure 7.39) shows the vortex, its feeding sheet and the shocks above and inboard of the vortex. The total pressure loss contours (Figure 7.40) show these features as well as the shock under the vortex. The leeward pressure is nearly constant for this case (Figures 7.41 and 7.42) and agrees well with the measured values. The cross-flow Mach number (Figures 7.43 and 7.44) is nearly constant on the windward side of the wing, but has large gradients on the leeward side. The feeding sheet and the three cross-flow shocks are shown clearly in the cross-flow, radial and total Mach numbers (Figures 7.44-7.46). The flow reaches a maximum cross-flow Mach number of 2.6 (before the cross-flow shock above the vortex), a maximum total Mach number of 9.5 (before the cross-flow shock beneath the vortex).

7.2. FLAT PLATE CASES

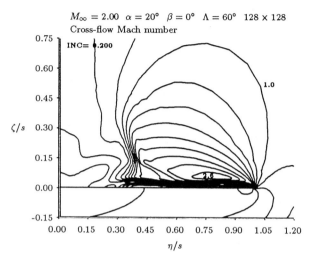

Figure 7.30: Cross-flow Mach number — $M_\infty = 2.0$, $\alpha = 20°$, $\Lambda = 60°$

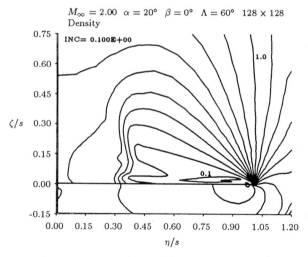

Figure 7.31: Density — $M_\infty = 2.0$, $\alpha = 20°$, $\Lambda = 60°$

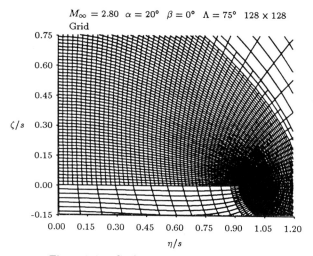

Figure 7.32: Grid — $M_\infty = 2.8$, $\alpha = 20°$, $\Lambda = 75°$

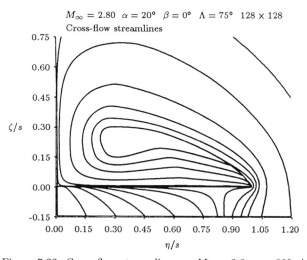

Figure 7.33: Cross-flow streamlines — $M_\infty = 2.8$, $\alpha = 20°$, $\Lambda = 75°$

7.2. FLAT PLATE CASES

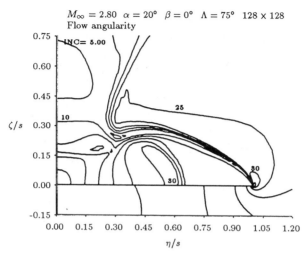

Figure 7.34: Flow angularity — $M_\infty = 2.8$, $\alpha = 20°$, $\Lambda = 75°$

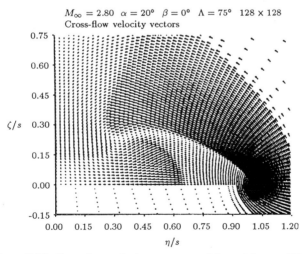

Figure 7.35: Cross-flow velocity vectors — $M_\infty = 2.8$, $\alpha = 20°$, $\Lambda = 75°$

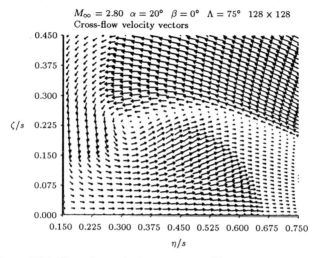

Figure 7.36: Cross-flow velocity vectors — $M_\infty = 2.8$, $\alpha = 20°$, $\Lambda = 75°$

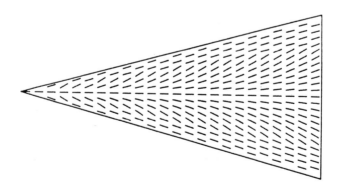

Figure 7.37: Computed tuft patterns — $M_\infty = 2.8$, $\alpha = 20°$, $\Lambda = 75°$

7.2. FLAT PLATE CASES

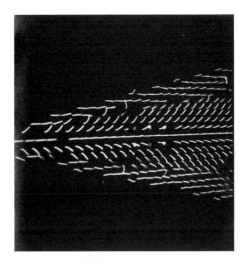

Figure 7.38: Experimental tuft patterns — $M_\infty = 2.8$, $\alpha = 20°$, $\Lambda = 75°$

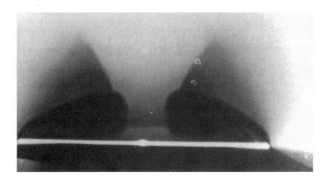

Figure 7.39: Experimental vapor screen — $M_\infty = 2.8$, $\alpha = 20°$, $\Lambda = 75°$

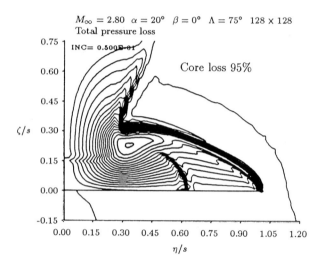

Figure 7.40: Total pressure loss — $M_\infty = 2.8$, $\alpha = 20°$, $\Lambda = 75°$

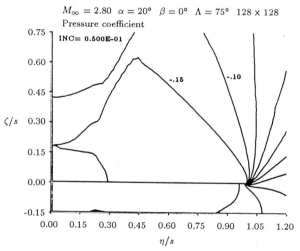

Figure 7.41: Pressure coefficient — $M_\infty = 2.8$, $\alpha = 20°$, $\Lambda = 75°$

7.2. FLAT PLATE CASES

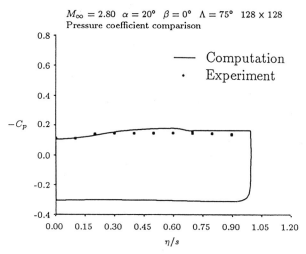

Figure 7.42: Pressure coefficient comparison — $M_\infty = 2.8$, $\alpha = 20°$, $\Lambda = 75°$

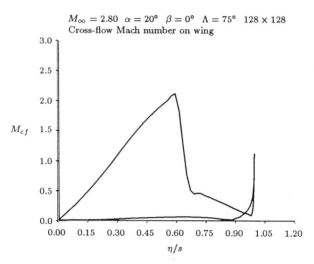

Figure 7.43: Cross-flow Mach number — $M_\infty = 2.8$, $\alpha = 20°$, $\Lambda = 75°$

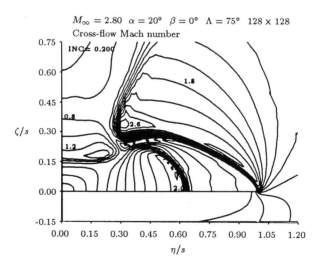

Figure 7.44: Cross-flow Mach number — $M_\infty = 2.8$, $\alpha = 20°$, $\Lambda = 75°$

7.2. FLAT PLATE CASES

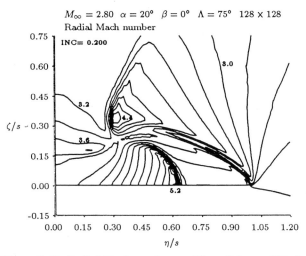

Figure 7.45: Radial Mach number — $M_\infty = 2.8$, $\alpha = 20°$, $\Lambda = 75°$

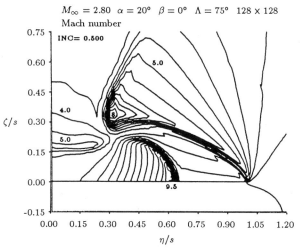

Figure 7.46: Mach number — $M_\infty = 2.8$, $\alpha = 20°$, $\Lambda = 75°$

7.3 Vortex flap cases

Vortex flaps are a drag-reduction idea for flows with leading-edge vortices. The concept is to locate the vortex on the deflected flaps, which have a forward facing normal, so that the low pressure region on the flap results in a thrust component, thereby reducing the drag (See Figure 7.47). The first experimental demonstration of the drag-reduction potential of vortex flaps took place at NASA Langley Research Center [51]. Computational results for vortex flap geometries have been reported by Luckring [31], Murman *et al* [40], Arlinger [3] and Powell *et al* [50]. This section presents the results for two vortex flap cases. They are:

1. $M_\infty = 2.4$, $\alpha = 4°$, $\delta = 10°$, 256×256 equivalent refinement;

2. $M_\infty = 2.4$, $\alpha = 12°$, $\delta = 10°$, 256×256 equivalent refinement.

The angle of attacks given above are the nominal angles of attack. After correction for sting deflection, the angles of attack were 3.84° and 12.68° respectively. The computations were carried out at the corrected angle of attack. The wing has a leading-edge sweep $\Lambda = 75°$ with the flap undeflected. The first case results in attached flow on the flap, separating at the hinge-line and rolling up into a vortex inboard of the hinge-line. The second case results in separation at the leading-edge, with a vortex sheet that is parallel to the flap, and a vortex inboard of the hinge-line.

7.3.1 Case 1 — $M_\infty = 2.4$, $\alpha = 4°$, $\delta = 10°$

This case was computed on a grid of 256×256 equivalent refinement, shown in Figure 7.48. The cross-flow streamlines (Figure 7.49) show the topology of the flow. The flow is attached at the leading-edge, separating at the hinge-line and rolling up into a vortex inboard of the hinge. There is a saddle point on the leeward side of the wing, dividing the cross-flow streamlines that roll up into the vortex from those that converge into the leeward symmetry point node. There is a saddle point at the leading-edge, dividing the windward and leeward portions of the flow. There is also a node at the windward symmetry point, into which all the cross-flow streamlines on the windward side converge. The angularity (Figure 7.50) is fairly low in this case; 10° above the vortex and 26° below it. The tuft patterns from the computation (Figure 7.51) show very low angularity everywhere but under the vortex.

7.3. VORTEX FLAP CASES

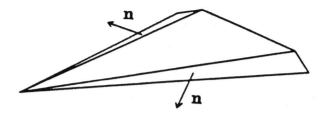

Figure 7.47: Vortex Flap Concept

under the vortex.

The vapor screen for this case (Figure 7.52) shows the vortex sheet (originating at the hinge line) and the primary vortex. The total pressure loss contours (Figure 7.53) show this same pattern. They also show a small loss on the windward side of the leading-edge. The cross-flow Mach number on the windward side of the wing (See Figure 7.54) increases from zero at the symmetry point to approximately 0.4 just inboard of the leading edge. There is a slight recompression outboard of this point. On the leeward side of the wing, there are three points where the cross-flow Mach number is approximately zero: the symmetry point, the separation at the hinge-line, and the saddle point inboard of the vortex. There is a weak expansion on the hinge, and a strong expansion, reaching supersonic speed, under the vortex. The contours of the cross-flow Mach number (Figure 7.55) show high gradients in the sheet and the core, and the expansion and recompression under the vortex.

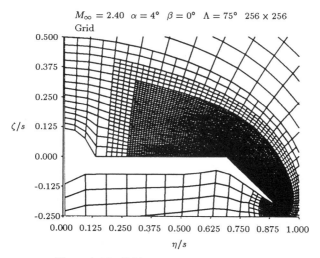

Figure 7.48: Grid — $M_\infty = 2.4$, $\alpha = 4°$, $\delta = 10°$

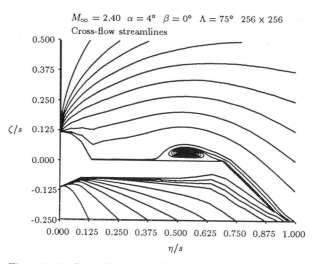

Figure 7.49: Cross-flow streamlines — $M_\infty = 2.4$, $\alpha = 4°$, $\delta = 10°$

7.3. VORTEX FLAP CASES

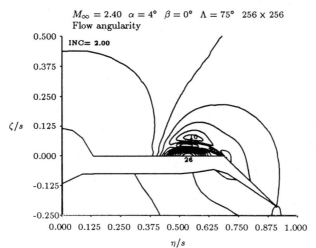

Figure 7.50: Flow angularity — $M_\infty = 2.4$, $\alpha = 4°$, $\delta = 10°$

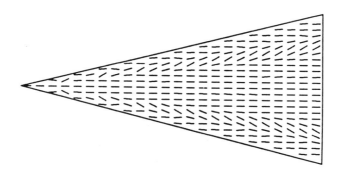

Figure 7.51: Computed tuft patterns — $M_\infty = 2.4$, $\alpha = 4°$, $\delta = 10°$

Figure 7.52: Vapor screen — $M_\infty = 2.4$, $\alpha = 4°$, $\delta = 10°$

The pressure coefficient contours (Figure 7.56) show that the pressure is nearly constant on the entire windward side, and on the leeward side inboard of the vortex. The expansion on the leeward side of the hinge and the expansion and recompression under the vortex are also shown. The pressure on the wing (Figure 7.57) compares well on the flap. Inboard of the vortex, the suction is under-predicted, suggesting a higher angle of attack in the experiment than in the computation. The suction peak under the vortex is over-predicted, and predicted too far outboard.

7.3.2   Case 2 — $M_\infty = 2.4$, $\alpha = 12°$, $\delta = 10°$

This case was run on an equivalent 256 × 256 grid, shown in Figure 7.58. The cross-flow streamlines (Figure 7.59) show a node at the windward symmetry point and saddle points at the leeward symmetry point, the leading-edge and the hinge-line. The leeward cross-flow streamlines all converge into the vortex; the windward streamlines into the windward symmetry point. The angularity (Figure 7.60) reaches 32° under the vortex and 20° above it. The high gradient of angularity above the flap suggests a cross-flow shock there. The computational tuft pattern (Figure 7.61) shows high angularity under the vortex, and moderate angularity on the flap.

The vapor screen for this case (Figure 7.62) shows the vortex and its feeding sheet. There is also evidence of a cross-flow shock and secondary vortex underneath the primary vortex, and a cross-flow shock above the flap. The contours of total pressure loss

7.3. VORTEX FLAP CASES

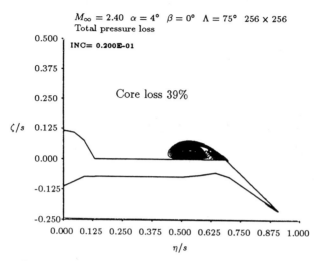

Figure 7.53: Total pressure loss — $M_\infty = 2.4$, $\alpha = 4°$, $\delta = 10°$

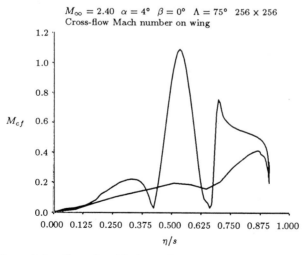

Figure 7.54: Cross-flow Mach number — $M_\infty = 2.4$, $\alpha = 4°$, $\delta = 10°$

154 CHAPTER 7. COMPARISON OF COMPUTATIONS AND EXPERIMENTS

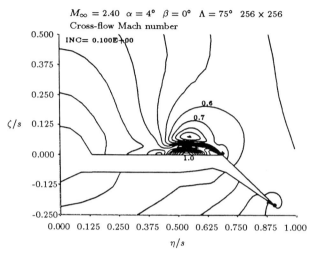

Figure 7.55: Cross-flow Mach number — $M_\infty = 2.4$, $\alpha = 4°$, $\delta = 10°$

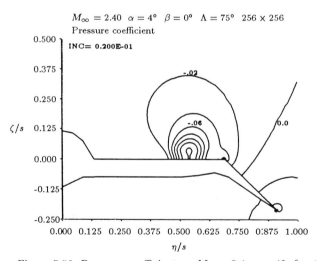

Figure 7.56: Pressure coefficient — $M_\infty = 2.4$, $\alpha = 4°$, $\delta = 10°$

7.3. VORTEX FLAP CASES

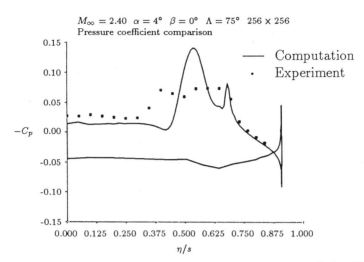

Figure 7.57: Pressure coefficient comparison — $M_\infty = 2.4$, $\alpha = 4°$, $\delta = 10°$

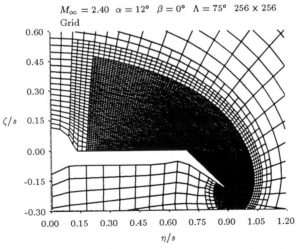

Figure 7.58: Grid — $M_\infty = 2.4$, $\alpha = 12°$, $\delta = 10°$

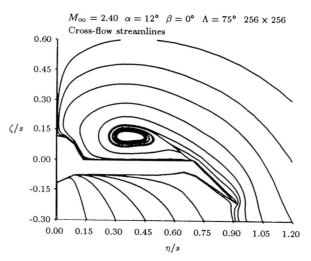

Figure 7.59: Cross-flow streamlines — $M_\infty = 2.4$, $\alpha = 12°$, $\delta = 10°$

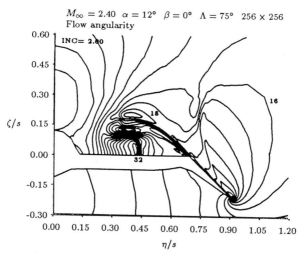

Figure 7.60: Flow angularity — $M_\infty = 2.4$, $\alpha = 12°$, $\delta = 10°$

7.3. VORTEX FLAP CASES

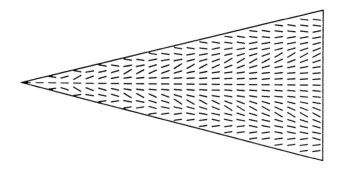

Figure 7.61: Computed tuft patterns — $M_\infty = 2.4$, $\alpha = 12°$, $\delta = 10°$

(Figure 7.63) show the sheet, the vortex and the two shocks. In this case, 98% of the total pressure is lost in the core of the vortex. The cross-flow Mach number is nearly constant on the windward side of the wing (See Figure 7.64). On the leeward side, the separation at the leading-edge, expansion and reattachment on the flap are shown. The flow then separates at the hinge-line. There is a rapid expansion and recompression *via* a cross-flow shock under the vortex. The cross-flow Mach number contours (Figure 7.65) show the feeding sheet and the two cross-flow shocks clearly. The Mach number and its radial component are shown in Figures 7.67 and 7.66. The density (Figure 7.68) reaches 0.3 in the core of the vortex.

The pressure gradients are fairly low in this case. The contours of pressure coefficient (Figure 7.69) show the low pressure region in the core, the cross-flow shocks above and below the vortex and the expansion and recompression on the flap. The pressure agrees very well with the experimental values (See Figure 7.70) except in the immediate vicinity of the vortex, where the suction due to the vortex is overpredicted.

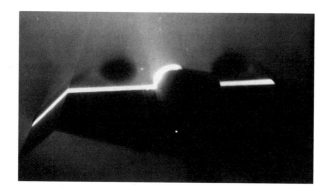

Figure 7.62: Vapor screen — $M_\infty = 2.4$, $\alpha = 12°$, $\delta = 10°$

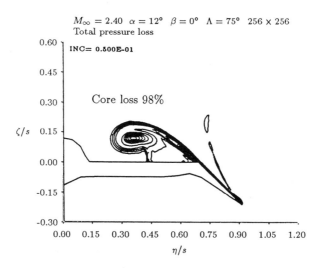

Figure 7.63: Total pressure loss — $M_\infty = 2.4$, $\alpha = 12°$, $\delta = 10°$

7.3. VORTEX FLAP CASES

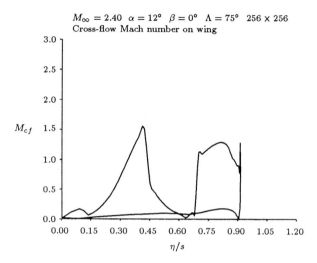

Figure 7.64: Cross-flow Mach number — $M_\infty = 2.4$, $\alpha = 12°$, $\delta = 10°$

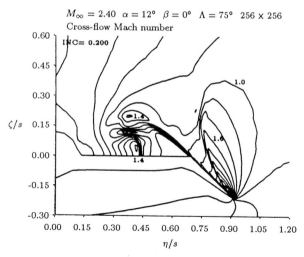

Figure 7.65: Cross-flow Mach number — $M_\infty = 2.4$, $\alpha = 12°$, $\delta = 10°$

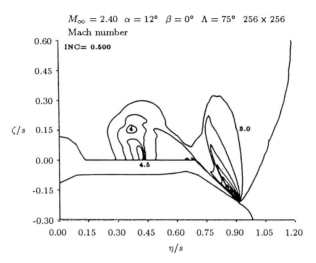

Figure 7.66: Radial Mach number — $M_\infty = 2.4$, $\alpha = 12°$, $\delta = 10°$

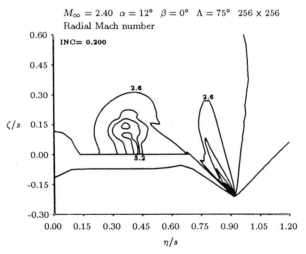

Figure 7.67: Mach number — $M_\infty = 2.4$, $\alpha = 12°$, $\delta = 10°$

7.3. VORTEX FLAP CASES

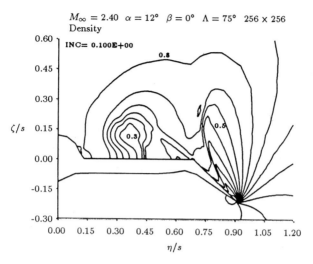

Figure 7.68: Density — $M_\infty = 2.4$, $\alpha = 12°$, $\delta = 10°$

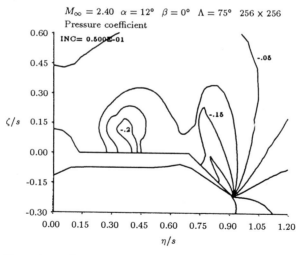

Figure 7.69: Pressure coefficient — $M_\infty = 2.4$, $\alpha = 12°$, $\delta = 10°$

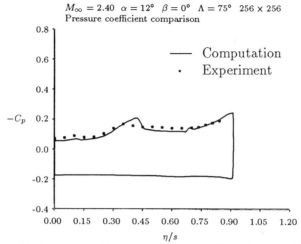

Figure 7.70: Pressure coefficient comparison — $M_\infty = 2.4$, $\alpha = 12°$, $\delta = 10°$

7.4 Asymmetric cases

This section presents the results for two yawed wing cases. They are

1. $M_\infty = 1.7$, $\alpha = 12°$, $\beta = 8°$, 256×128 equivalent refinement;

2. $M_\infty = 2.8$, $\alpha = 12°$, $\beta = 8°$, 256×128 equivalent refinement.

The angles of attack with sting deflection accounted for were 11.72° and 11.85°. The computations were carried out at the corrected angles of attack. Both wings have a leading-edge sweep $\Lambda = 75°$. The first case has vortices on both the port and starboard sides of the wing; the second has attached flow on the port side, with a strong cross-flow shock near the center-line of the wing.

7.4.1 Case 1 — $M_\infty = 1.7$, $\alpha = 12°$, $\beta = 8°$

This case was run on a grid with 256×128 equivalent refinement, shown in Figure 7.71. The cross-flow streamline plot (Figure 7.72) shows two vortices: a long, narrow vortex on the port side and a nearly circular vortex on the starboard side of the wing. There is a saddle point on the windward side of the wing near the port leading-edge, and one on the leeward side near the center of the wing. All the cross-flow streamlines to port of the two saddle points converge in the port vortex; all others converge in the starboard

7.4. ASYMMETRIC CASES

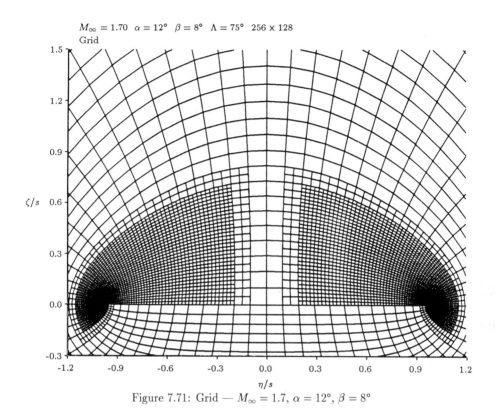

Figure 7.71: Grid — $M_\infty = 1.7$, $\alpha = 12°$, $\beta = 8°$

vortex. The flow angularity (Figure 7.73) is high above and below both vortices, and low elsewhere in the flow. The tuft patterns (Figure 7.74) show the extent of the port and starboard vortices.

The total pressure loss contours (Figure 7.75) show the difference in strength and shape of the two vortices. The port vortex is stronger, with a core total pressure loss of 66%. The starboard vortex has a core loss of 43%. There is also evidence of a weak shock above the port vortex. The pressure coefficient contours (Figure 7.76) also show evidence of this shock. The pressure on the wing (Figure 7.77) compares well with experiment. Both vortices are predicted too far outboard, and the suction peaks are overpredicted. The solution does capture the unusual double expansion-recompression pattern under the port vortex, however.

164 CHAPTER 7. COMPARISON OF COMPUTATIONS AND EXPERIMENTS

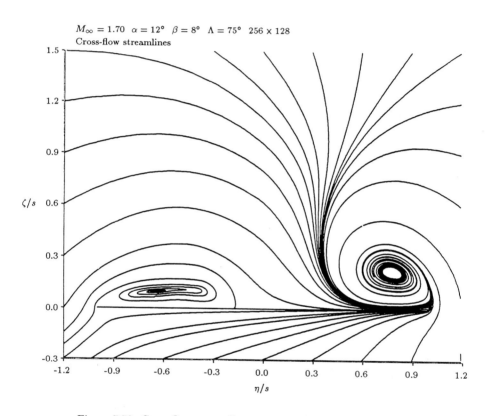

Figure 7.72: Cross-flow streamlines — $M_\infty = 1.7$, $\alpha = 12°$, $\beta = 8°$

7.4. ASYMMETRIC CASES

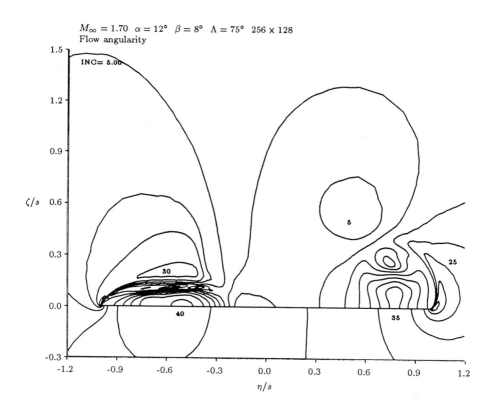

Figure 7.73: Flow angularity — $M_\infty = 1.7$, $\alpha = 12°$, $\beta = 8°$

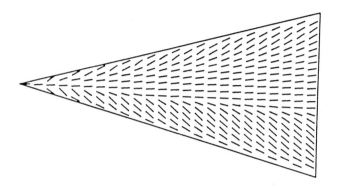

Figure 7.74: Computed tuft patterns — $M_\infty = 1.7$, $\alpha = 12°$, $\beta = 8°$

This unusual pattern is more clearly seen in the cross-flow Mach number distribution (Figures 7.78 and 7.79). Under the port vortex, the flow expands to a cross-flow Mach number of 1.6, recompresses to 1.2, then expands again, reaching a value of 1.3, then recompresses outboard to the port leading-edge. The cross-flow Mach number reaches a value of 1.7 above the port vortex, and goes slightly supersonic above and below the starboard vortex.

7.4.2 Case 2 — $M_\infty = 2.8$, $\alpha = 12°$, $\beta = 8°$

The grid for this case is shown in Figure 7.80. It has equivalent 256×128 refinement. The cross-flow streamlines for this case show attached flow on the port side of the wing and a vortex on the starboard side. There is a saddle point at the port leading-edge, separating the windward and leeward cross-flow streamlines. All of the streamlines converge in the starboard vortex for this case. The vortex has extremely small inflow velocities; the trajectory integration reached a limit cycle even for very small values of Δs. The flow angularity contours show high gradients at the shock under the starboard vortex, in the feeding sheet for the starboard vortex, at the shock on the leeward side, just to port of the center of the wing, and through the bow shock near the port leading-edge. The tuft

7.4. ASYMMETRIC CASES

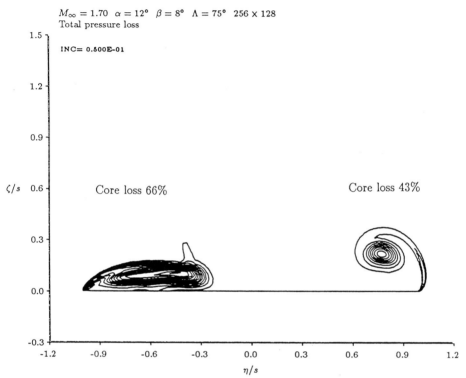

Figure 7.75: Total pressure loss — $M_\infty = 1.7$, $\alpha = 12°$, $\beta = 8°$

168   CHAPTER 7. COMPARISON OF COMPUTATIONS AND EXPERIMENTS

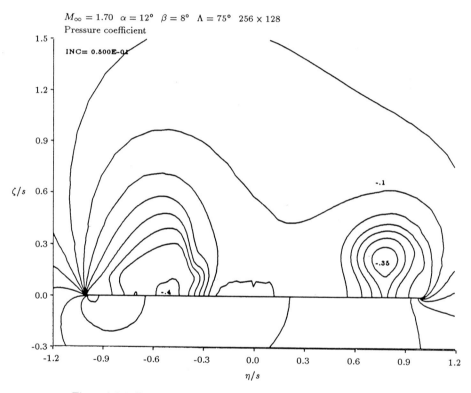

Figure 7.76: Pressure coefficient — $M_\infty = 1.7$, $\alpha = 12°$, $\beta = 8°$

7.4. ASYMMETRIC CASES

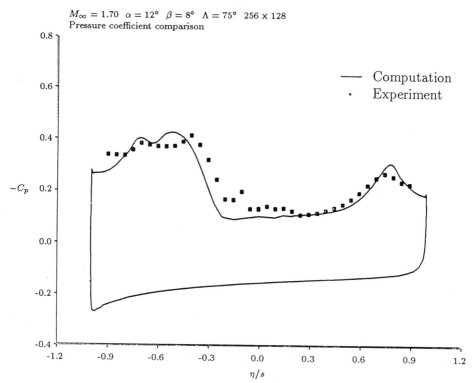

Figure 7.77: Pressure coefficient comparison — $M_\infty = 1.7$, $\alpha = 12°$, $\beta = 8°$

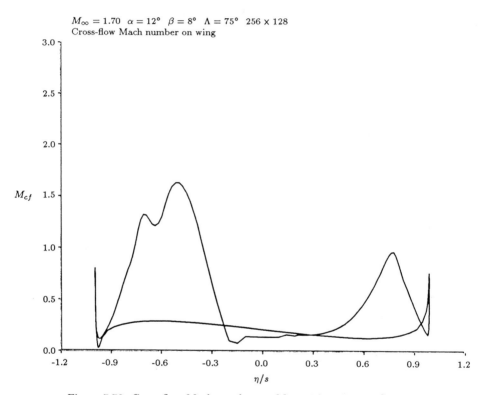

Figure 7.78: Cross-flow Mach number — $M_\infty = 1.7$, $\alpha = 12°$, $\beta = 8°$

7.4. ASYMMETRIC CASES

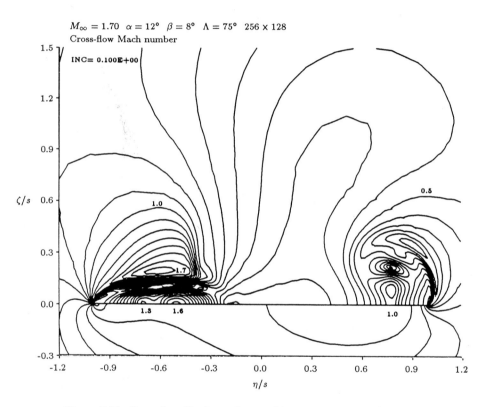

Figure 7.79: Cross-flow Mach number — $M_\infty = 1.7$, $\alpha = 12°$, $\beta = 8°$

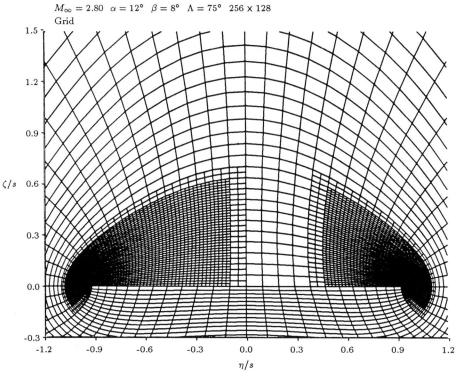

Figure 7.80: Grid — $M_\infty = 2.8$, $\alpha = 12°$, $\beta = 8°$

plot (Figure 7.83) shows high angularity on the entire port side of the wing and on the outboard portion of the leeward side of the wing.

The total pressure loss contours (Figure 7.84) show a large loss at the port leading-edge which is convected to the cross-flow shock near the centerline of the wing. The starboard shock-vortex system, with a core loss of 65%, is also shown. The pressure coefficient contours (Figure 7.85) show the bow shock near the port leading-edge, the leading-edge expansions and the low pressure in the core of the starboard vortex. The pressure gradient is fairly low on the leeward port side. The comparison with experiment (Figure 7.86) is good. The suction peak of the starboard vortex is over-predicted, and cross-flow shock on the port side is predicted too far outboard. The pressures near the centerline do not agree well, due to the centerbody which was used in the experiment and not modeled in the computation.

7.4. ASYMMETRIC CASES

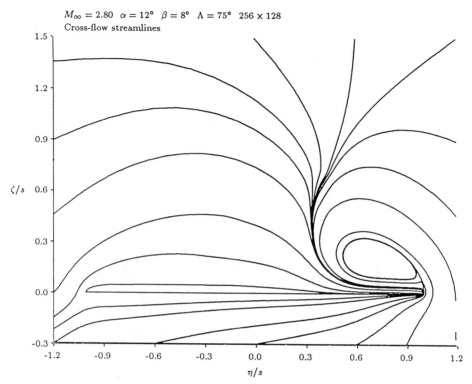

Figure 7.81: Cross-flow streamlines — $M_\infty = 2.8$, $\alpha = 12°$, $\beta = 8°$

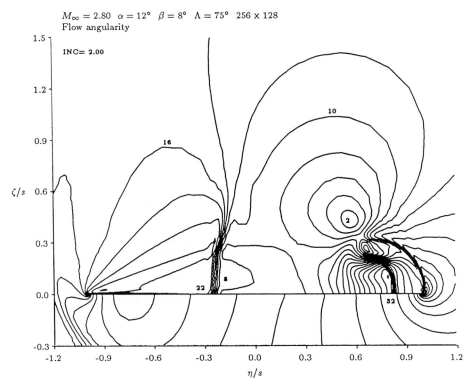

Figure 7.82: Flow angularity — $M_\infty = 2.8$, $\alpha = 12°$, $\beta = 8°$

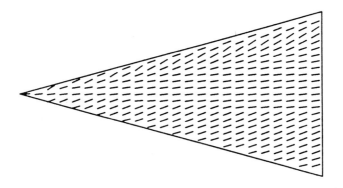

Figure 7.83: Computed tuft patterns — $M_\infty = 2.8$, $\alpha = 12°$, $\beta = 8°$

The cross-flow Mach number on the wing (Figure 7.87) is low on the windward side of the wing. The rapid expansion at the port leading-edge is shown, and the strong cross-flow shock near the center-line. The starboard vortex has a strong expansion underneath it, with recompression *via* a cross-flow shock. The cross-flow Mach number contours (Figure 7.88) show the bow shock near the port leading-edge, the strong cross-flow shock near the center-line and the starboard vortex, feeding-sheet and shock.

7.5 Summary

Experimental and numerical results have been presented for three symmetric flat plate cases, two vortex flap cases and two asymmetric flat plat cases. A variety of flow topologies have been represented, with the computation correctly predicting the gross features of the flow in each case. The pressures on the wing agree very well on the flat plate cases, although the computations slightly overpredict the suction peak under the primary vortices and do not model the secondary vortices. The pressures also agree well in the high Mach number vortex flap case. In the low Mach number vortex flap case, the suction

176 CHAPTER 7. COMPARISON OF COMPUTATIONS AND EXPERIMENTS

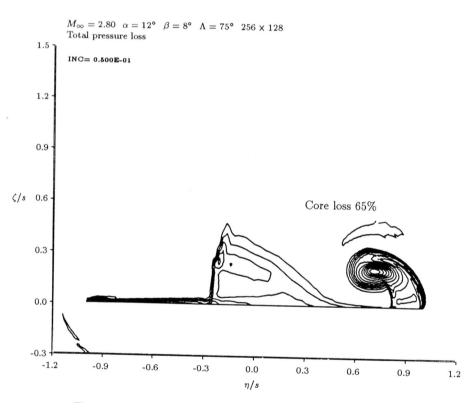

Figure 7.84: Total pressure loss — $M_\infty = 2.8$, $\alpha = 12°$, $\beta = 8°$

7.5. SUMMARY

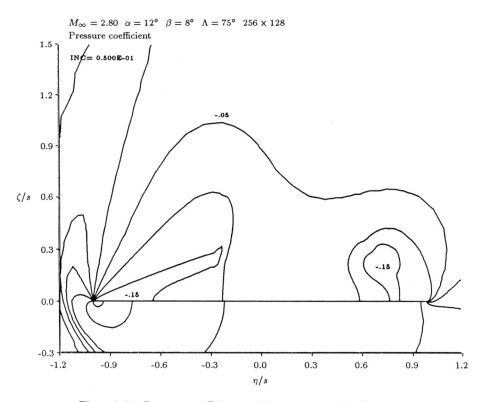

Figure 7.85: Pressure coefficient — $M_\infty = 2.8$, $\alpha = 12°$, $\beta = 8°$

178   CHAPTER 7. COMPARISON OF COMPUTATIONS AND EXPERIMENTS

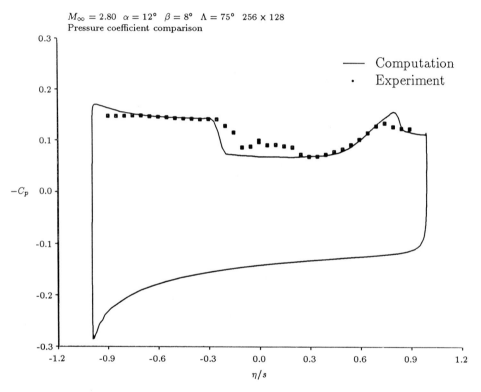

Figure 7.86: Pressure coefficient comparison — $M_\infty = 2.8$, $\alpha = 12°$, $\beta = 8°$

7.5. SUMMARY

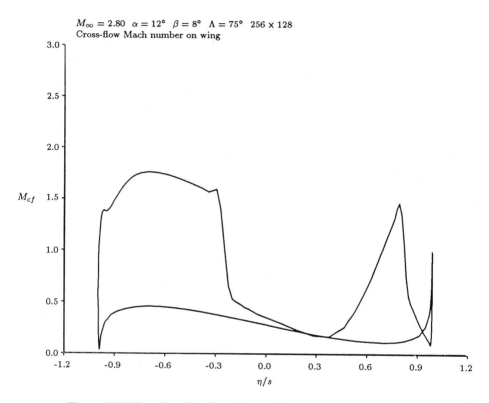

Figure 7.87: Cross-flow Mach number — $M_\infty = 2.8$, $\alpha = 12°$, $\beta = 8°$

180 CHAPTER 7. COMPARISON OF COMPUTATIONS AND EXPERIMENTS

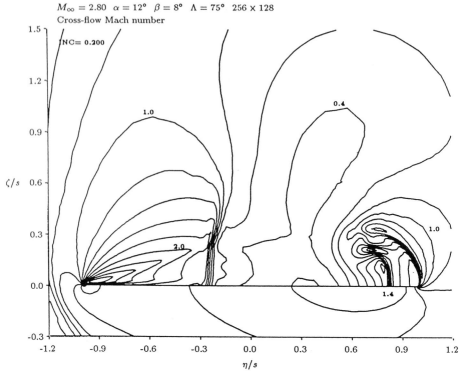

Figure 7.88: Cross-flow Mach number — $M_\infty = 2.8$, $\alpha = 12°$, $\beta = 8°$

7.1. SUMMARY

peak due to the vortex inboard of the hinge-line is overpredicted. In both of the vortex flap cases, the flow leaves the hinge-line parallel to the vortex flap, suggesting that a Kutta condition is being enforced implicitly at the hinge-line. This result has also been seen in a more comprehensive study of Euler calculations for delta wings with vortex flaps [50].

Chapter 8

Conclusions

There have been three primary goals of the research presented in this book, all directed towards understanding leading-edge vortex flows. The first has been the development of an algorithm to produce solutions for these flows. The governing equations, the conical Euler equations, have been presented in Chapter 2, along with some of the repercussions of the choice of the model. The solution algorithm, a new cell-vertex scheme which allows for multiply-embedded regions, has been presented in Chapter 3.

The second goal has been to understand the solutions produced by this algorithm, with particular emphasis on the total pressure losses. Chapter 4 has presented results from two cases, explaining the trends in the different flow variables. Chapter 5 has presented an exhaustive study of the total pressure losses that occur in the solutions, and how they are affected by different parameters of the computation. Chapter 6 has put forward an explanation for the behavior of the losses.

The final goal has been to present numerical and experimental results for a number of cases, and use them to shed some light on the physics of these flows. Results from seven cases have been presented, and compared with experiment. The outcome of the work towards the three goals is summarized below, and recommendations for further research are given.

8.1 Algorithm Development

The governing equation set chosen was the conical Euler equations. The use of this model made the computations inexpensive, relative to the full three-dimensional model. This made it possible to run a large number of cases, only a small fraction of which are reported here. This greatly helped in the understanding of the problem.

The solution algorithm, presented in Chapter 3, is a new cell-vertex method for grids with multiply-embedded regions. It is comprised of a second-order accurate finite-volume discretization, a multi-stage time integration and an artificial viscosity composed of a linear fourth difference and nonlinear second difference. A method of handling the embedding interfaces in the flux and damping calculations was presented, along with a data structure for the scheme. The scheme proved robust and efficient. The use of embedded regions provided savings of a factor of up to fifty in computer time.

8.2 Characteristics of the Solutions

The basic characteristics of the conical Euler solutions were studied. Regions of low pressure and density were seen in the vortex. Rapid expansions were seen at the leading-edge and underneath the vortex. Gradients in the cross-flow Mach number were very high on the leeward side of the wing and low on the windward side. Gradients in the axial Mach number were much lower. Large total pressure losses in the region of the vortex were consistently seen. A study of the effects of numerical parameters such as grid resolution and artificial viscosity showed that the level of the total pressure loss was independent of numerical parameters, while the distribution of the loss was not. Two scales for the size of the vortex were presented: a macro-scale, which provided an estimate of the size of the leading-edge vortex; and a micro-scale, which provided an estimate of the size of the viscous sub-core. The macro-scale was found to be relatively insensitive to numerical parameters; the micro-scale was found to be extremely sensitive. The level of the total pressure loss was found to depend upon physical parameters such as the angle of attack, free-stream Mach number and leading-edge sweep. It also was found to depend upon the form of the artificial viscosity used.

A study was made of the level to which the Euler equations were being satisfied in the vortex core region. An equivalent Reynolds number was calculated, and found to be on the order of one hundred on coarse grids, and one thousand on fine grids. The error in Crocco's relation was calculated, and found to be order one in the core on all grids. The total pressure losses were compared with those seen experimentally by presenting computed pitot pressures and experimental measurements of pitot pressures. The losses were found to be very realistic in the primary vortex. Lossless solutions were calculated by replacing the x momentum equation with a constant total pressure condition. The two solutions differed only in the core of the vortex, where the non-isentropic solutions are more realistic than the lossless solutions.

The computational study of the total pressure losses suggested that the level of artificial viscosity in the vortex was not negligible. A model for the losses was proposed, based on this conclusion. The model for the feeding sheet was based on the fact that the contact discontinuity weak solution is degenerate, and not well modeled by the discrete equations. The sheet is given a structure, with an associated loss. The model for the core was based on a new similarity solution to the incompressible, axisymmetric, conical Navier-Stokes equations — a high Reynolds number, slender core limit. This similarity solution showed the total pressure loss level in a vortex to be independent of Reynolds number, depending solely on the edge swirl velocity of the vortex.

8.3 Comparison of Numerical and Physical Results

Results were presented for three symmetric flat plate cases, two vortex flap cases, and two asymmetric flat plate cases. All were compared to experimental data obtained from NASA Langley Research Center. The comparisons were remarkably good in the symmetric and asymmetric flat plate cases. The vortex flap cases were not as good, due to viscous effects at the hinge-line. In all of the cases, the computations provided a good model for the topology of the flow. Suction peaks under the vortices tended to be over-predicted and placed too far outboard.

8.4 Recommendations for Further Research

1. The first item to be addressed is the model chosen. The conical Euler equations provide an inexpensive way of understanding the gross physics of these flows. They are restricted, however, in their inability to model boundary layers, secondary vortices and three-dimensional effects. An extension of the scheme to conical Navier-Stokes is fairly straightforward. The extension to three-dimensional Euler and Navier-

Stokes is less so, but certainly possible. The extension is necessary to investigate flows which are not approximately conical.

2. The second item to be addressed is the solution algorithm. Although the scheme proved efficient and robust for the cases given, a number of improvements could be made. High Mach number cases with cross-flow shocks make flux-vector or flux-difference splitting techniques attractive. The increase in efficiency that was gained by allowing for rectangular embedded regions could be doubled by allowing for more general embedded regions whose locations are found adaptively. The extension to more general regions is very little work — the only change necessary would be a slight generalization of the interface pointers. It would remain to choose an adaptation parameter (or parameters) that would flag cells in regions where refinement is desired.

3. The third item to be addressed is the core model. For direct comparison with the solutions, the model should be extended, if possible, to compressible flow. It would also be interesting to investigate the extent to which the insight from the Burgers and conical vortex models applies to other, less restrictive vortex flows. For this, numerical solutions of the Navier-Stokes equations may be needed. The core model also suggests that experiments should be performed to verify that the total pressure loss and swirl angle are directly related and are Reynolds number independent. Finally, the effects of turbulence on the core model should be investigated.

Bibliography

[1] S. R. Allmaras. Embedded Mesh Solutions of the 2-D Euler Equations Using a Cell-Centered Finite Volume Scheme. Master's thesis, Massachusetts Institute of Technology, 1985.

[2] D. A. Anderson, J. C. Tannehill, and R. H. Pletcher. *Computational Fluid Mechanics and Heat Transfer*. Hemisphere Publishing Corporation, 1984.

[3] B. G. Arlinger. Computation of Supersonic Flow Including Leading-Edge Vortex Flows Using Marching Euler Technique. Presented at the International Symposium on Computational Fluid Dynamics, Tokyo, Japan, September 1985.

[4] G. R. Baker. The "Cloud in Cell" Technique Applied to the Roll up of Vortex Sheets. *Journal of Computational Physics*, volume 31, pages 76–95, 1979.

[5] A. L. Braslow and E. C. Knox. Simplified Method for Determination of Critical Height of Distributed Roughness Particles for Boundary-Layer Transition at Mach Numbers from 0 to 5. TN 4363, NACA, 1958.

[6] C. E. Brown and W. H. Michael. Effect of Leading-Edge Separation on the Lift of a Delta Wing. *Journal of Aeronautical Sciences*, volume 21, pages 690–694, 1954.

[7] S. N. Brown. The Compressible Inviscid Leading-Edge Vortex. *Journal of Fluid Mechanics*, volume 22, pages 17–32, 1965.

[8] S. N. Brown and K. W. Mangler. An Asymptotic Solution for the Centre of a Rolled-up Conical Vortex Sheet in Compressible Flow. *The Aeronautical Quarterly*, pages 354–366, 1967.

[9] J. M. Burgers. A Mathematical Model Illustrating the Theory of Turbulence. In *Advances in Applied Mechanics*, volume 1, pages 197–199. Academic Press, 1948.

[10] S. R. Chakravarthy and D. K. Ota. Numerical Issues in Computing Inviscid Supersonic Flow Over Conical Delta Wings. AIAA Paper 86-0440, January 1986.

[11] A. J. Chorin and P. S. Bernard. Discretization of a Vortex Sheet, with an Example of Roll-up. *Journal of Computational Physics*, volume 13, pages 423–429, 1973.

[12] J. F. Dannenhoffer III and J. R. Baron. Robust Grid Adaptation for Complex Transonic Flows. AIAA Paper 86-0495, 1986.

[13] A. Eberle, A. Rizzi, and E. H. Hirschel. *Numerical Solutions of the Euler Equations for Steady Flow Problems. Notes on Numerical Fluid Mechanics*, Vieweg, 1990. To appear.

[14] L. E. Eriksson. *Transfinite Mesh Generation and Computer-Aided Analysis of Mesh Effects*. PhD thesis, Uppsala University, 1984.

[15] A. Ferri. *Supersonic Flows with Shock Waves*, chapter H. General Theory of High-speed Aerodynamics. OUP, 1955.

[16] K. Fujii and P. Kutler. Numerical Simulation of the Viscous Flow Fields over Three-Dimensional Complicated Geometries. AIAA Paper 84-1550, June 1984.

[17] M. B. Giles. Energy Stability Analysis of Multi-Step Methods on Unstructured Meshes. TR 87-1, MIT-CFDL, March 1987.

[18] J. P. Guiraud and R. Kh. Zeytounian. A double-scale investigation of the asymptotic structure of rolled-up vortex sheets. *Journal of Fluid Mechanics*, volume 79, pages 93–112, 1977.

[19] M. G. Hall. A Theory for the Core of a Leading-Edge Vortex. *Journal of Fluid Mechanics*, volume 11, pages 209–228, 1961.

[20] M.G. Hall. Cell-Vertex Schemes for Solution of the Euler Equations. Tech. Memo Aero 2029, Royal Aircraft Establishment, March 1985.

[21] E. H. Hirschel and A. Rizzi. The Mechanism of Vorticity creation in Euler Solutions for Lifting Wings. In *Proceedings of the Symposium on International Vortex Flow Experiment on Euler Code Validation*, ISBN 91-97 0914-0-5, 1986.

BIBLIOGRAPHY

[22] H. W. M. Hoeijmakers. Numerical Simulation of Vortical Flow. MP 86032 U, NLR, 1986.

[23] C. M. Jackson, W. A. Corlett, and W. J. Monta. Description and Calibration of the Langley Unitary Wind Tunnel. TP 1905, NASA, 1981.

[24] A. Jameson, W. Schmidt, and E. Turkel. Numerical Solutions of the Euler Equations by a Finite Volume Method Using Runge-Kutta Time-Stepping Schemes. AIAA Paper 81-1259, June 1981.

[25] F. T. Johnson, E. N. Tinoco, P. Lu, and M. A. Epton. Three-Dimensional Flow over Wings with Leading-Edge Vortex Separation. *AIAA Journal*, volume 18, number 4, pages 367–380, April 1980.

[26] J. G. Kallinderis and J. R. Baron. Adaptation Methods for a New Navier-Stokes Algorithm. AIAA Paper 87-1167-CP, 187.

[27] O. A. Kandil and A. Chuang. Influence of Numerical Dissipation in Computing Supersonic Vortex-Dominated Flows. AIAA paper 86-1073, May 1986.

[28] H. B. Keller. A New Difference Scheme for Parabolic Problems. In *Numerical Solutions of Partial Differential Equations*, volume 2. Academic Press, 1970.

[29] D. Küchemann. Report on the IUTAM Symposium on Concentrated Vortex Motions in Fluids. *Journal of Fluid Mechanics*, volume 21, number 1, pages 1–20, 1965.

[30] H. W. Liepmann and A. Roshko. *Elements of Gasdynamics*. John Wiley & Sons, 1957.

[31] J. M. Luckring. A Theory for the Core of a Three-Dimensional Leading-Edge Vortex. AIAA Paper 85-0108, January 1985.

[32] K. W. Mangler and J. H. B. Smith. A Theory of the Flow past a Slender Delta Wing with Leading-Edge Separation. *Proceedings of the Royal Society of London Series A*, volume 251, pages 200–217, 1959.

[33] K. W. Mangler and J. Weber. The Flow Field near the Centre of a Rolled-Up Vortex Sheet. *Journal of Fluid Mechanics*, volume 30, pages 177–196, 1967.

[34] F. Marconi. The Spiral Singularity in the Supersonic Inviscid Flow over a Cone. AIAA Paper 83-1665, July 1983.

[35] S. N. McMillin, J. L. Thomas, and E. M. Murman. Euler and Navier-Stokes Solutions for the Leeside Flow over Delta Wings at Supersonic Speeds. AIAA Paper 87-2270-CP, August 1987.

[36] D. S. Miller and R. M. Wood. Lee-Side Flow Over Delta Wings at Supersonic Speeds. TP 2430, NASA, June 1985.

[37] B. Monnerie and H. Werlé. Étude de l'Écoulement Supersonique & Hypersonique autour d'une Aile Élancée en Incidence. In *AGARD-CP-30*, 1968. Paper 23.

[38] D. W. Moore. On the Point Vortex Method. *SIAM Journal of Scientific and Statistical Computing*, volume 2, number 1, pages 65–84, 1981.

[39] B. Müller and A. Rizzi. Navier-Stokes Computation of Transonic Vortices over a Round Leading Edge Delta Wing. AIAA Paper 87-1227, June 1987.

[40] E. M. Murman, K. G. Powell, D. S. Miller, and R. M. Wood. Comparison of Computations and Experimental Data for Leading-Edge Vortices - Effects of Yaw and Vortex Flaps. AIAA Paper 86-0439, January 1986.

[41] E. M. Murman, A. Rizzi, and K. G. Powell. High Resolution Solutions of the Euler Equations for Vortex Flows. In *Progress and Supercomputing in Computational Fluid Dynamics*, pages 93–113. Birkhauser-Boston, 1985.

[42] E. M. Murman and P. M. Stremel. A Vortex Wake Capturing Method for Potential Flow Calculations. AIAA paper 82-0947, 1982.

[43] R. W. Newsome. A Comparison of Euler and Navier-Stokes Solutions for Supersonic Flow over a Conical Delta Wing. AIAA Paper 85-0111, January 1985.

[44] R. W. Newsome and O. A. Kandil. Vortical Flow Aerodynamics - Physical Aspects and Numerical Simulation. AIAA Paper 87-0205, January 1987.

[45] R. W. Newsome and J. L. Thomas. Computation of Leading-Edge Vortex Flows. CP 2416, NASA, 1985.

BIBLIOGRAPHY

[46] R. H. Ni. A Multiple-Grid Scheme for Solving the Euler Equations. *AIAA Journal*, volume 20, number 11, pages 1565–1571, 1981.

[47] E. C. Polhamus. A Concept of the Vortex Lift of Sharp-Edge Delta Wings Based on a Leading-Edge Suction Analogy. TN D-3767, NASA, 1966.

[48] K. G. Powell and E. M. Murman. Vortical Solutions of the Conical Euler Equations. Presented at the SIAM 1986 National Meeting, Boston, MA, July 1986.

[49] K. G. Powell, E. M. Murman, E. S. Perez, and J. R. Baron. Total Pressure Loss in Vortical Solutions of the Conical Euler Equations. *AIAA Journal*, volume 25, number 3, pages 360–368, March 1987.

[50] K. G. Powell, E. M. Murman, R. M. Wood, and D. S. Miller. A Comparison of Experimental and Numerical Results for Delta Wings with Vortex Flaps. AIAA Paper 86-1840-CP, June 1986.

[51] D. M. Rao. Leading-Edge Vortex Flap Experiments on a 74-Degree Delta Wing. CR 159161, NASA, 1979.

[52] R.D. Richtmyer. *Principles of Advanced Mathematical Physics*, volume 1. Springer-Verlag, 1978.

[53] D. P. Rizetta and J. S. Shang. Numerical Simulation of Leading-Edge Vortex Flows. AIAA Paper 84-1544, June 1984.

[54] A. Rizzi. Three-Dimensional Solutions to the Euler Equations with One Million Grid Points. *AIAA Journal*, volume 23, number 12, page 1986, December 1985.

[55] A. Rizzi and L. E. Eriksson. Computation of Flow arouna Wings Based on the Euler Equations. *Journal of Fluid Mechanics*, volume 148, pages 45–72, November 1984.

[56] L. Rosenhead. The Formation of Vortices from a Surface of Discontinuity. *Proceedings of the Royal Society of London Series A*, volume 134, pages 170–192, 1931.

[57] R. S. Shapiro and E. M. Murman. Cartesian Grid Finite Element Solutions to the Euler Equations. AIAA Paper 87-0559, 1987.

[58] G. F. Simmons. *Differential Equations with Applications and Historical Notes*. McGraw-Hill, 1972.

[59] A. Stanbrook and L. C. Squire. Possible Types of Flow at Swept Leading Edges. *Aeronautical Quarterly*, volume 15, number 1, pages 72–82, February 1964.

[60] J. L. Steger and R. L. Sorenson. Automatic Mesh-Point Clustering Near a Boundary in Grid Generation with Elliptic Partial Differential Equations. *Journal of Computational Physics*, volume 33, number 3, pages 405–410, December 1979.

[61] K. Stewartson and M. G. Hall. The Inner Viscous Solution for the Core of a Leading-Edge Vortex. *Journal of Fluid Mechanics*, volume 15, pages 306–318, 1963.

[62] J. Szodruch and D. Peake. Leeward Flow over Delta Wings at Supersonic Speeds. TM 81187, NASA, 1981.

[63] J. L. Thomas, S. L. Taylor, and K. Anderson. Navier-Stokes Computations of Vortical Flows over Low Aspect Wings. AIAA Paper 87-207, January 1987.

[64] E. Turkel. Accuracy of Schemes with Nonuniform Meshes for Compressible Fluid Flows. Technical Report 85-43, ICASE, 1985.

[65] W. J. Usab. *Embedded Mesh Solution of the Euler Equations Using a Multiple-Grid Method*. PhD thesis, Massachusetts Institute of Technology, 1984.

[66] Y. C. Vigneron, J. V. Rakich, and J. C. Tannehill. Calculation of Supersonic Viscous Flow over Delta Wings with Sharp Subsonic Leading Edges. AIAA Paper 78-1137, 1978.

[67] G. Vorropoulos and J. F. Wendt. Laser Velocimetry Study of Compressibility Effects on the Flow Field of a Delta Wing. In *AGARD-CP-342*, April 1983. Paper 9.

[68] N. J. Zabusky, M. H. Hughes, and K. V. Roberts. Contour Dynamics for the Euler Equations in Two Dimensions. *Journal of Computational Physics*, volume 30, pages 96–106, 1979.

Appendix A

Stability Analysis

A scalar model equation for the conical Euler equations is

$$\frac{\partial \tilde{u}}{\partial t} + \frac{\partial \tilde{u}}{\partial x} + \frac{\partial \tilde{u}}{\partial y} + a\tilde{u} = \mu_2 \left(\frac{\partial^2 \tilde{u}}{\partial x^2} + \frac{\partial^2 \tilde{u}}{\partial y^2} \right) - \mu_4 \left(\frac{\partial^4 \tilde{u}}{\partial x^4} + \frac{\partial^4 \tilde{u}}{\partial y^4} \right), \tag{A.1}$$

where a is a positive constant used to model the source term. The source term simply acts as a stabilizing factor however, and may be removed by setting

$$u(x,y,t) = e^{at}\tilde{u}(x,y,t), \tag{A.2}$$

which gives

$$\frac{\partial u}{\partial t} + \frac{\partial u}{\partial x} + \frac{\partial u}{\partial y} = \mu_2 \left(\frac{\partial^2 u}{\partial x^2} + \frac{\partial^2 u}{\partial y^2} \right) - \mu_4 \left(\frac{\partial^4 u}{\partial x^4} + \frac{\partial^4 u}{\partial y^4} \right). \tag{A.3}$$

The semi-discrete analysis is carried out by setting

$$u(x,y,t) = \hat{u}(t) e^{ik_x \Delta x} e^{ik_y \Delta y}. \tag{A.4}$$

The trapezoidal integration over a cell of sides Δx and Δy gives the convective fluxes

$$\begin{aligned}\sum F &= -\frac{1}{2}(u_{SW} + u_{SE})\Delta x + \frac{1}{2}(u_{SE} + u_{NE})\Delta y + \\ &+ \frac{1}{2}(u_{NE} + u_{NW})\Delta x - \frac{1}{2}(u_{NW} + u_{SW})\Delta y = \\ &= (u_{NE} - u_{SW})\frac{\Delta x + \Delta y}{2} + (u_{NW} - u_{SE})\frac{\Delta x - \Delta y}{2}.\end{aligned} \tag{A.5}$$

The flux is distributed to a node with a $(\frac{1}{4}, \frac{1}{4}, \frac{1}{4}, \frac{1}{4})$ distribution, giving, upon substitution of Equation A.4

$$Ru = \frac{i}{4}\{[1+ Æ\,](\sin[k_x + k_y\,Æ\,]\Delta + \sin[k_y\,Æ\,\Delta] + \sin[k_x\Delta]) +$$
$$+ [1-Æ\,](\sin[k_y\,Æ\,-k_x]\Delta + \sin[k_y\,Æ\,\Delta] - \sin[k_x\Delta])\}\,\hat{u}, \quad (A.6)$$

where

$$Æ = \frac{\Delta y}{\Delta x} = \frac{\Delta y}{\Delta}. \quad (A.7)$$

The viscous operators are

$$D_2 u = \{-2(1-\cos[k_x + k_y\,Æ\,]\Delta) - 2(1-\cos[k_x - k_y\,Æ\,]\Delta) -$$
$$- 4(1-\cos[k_x\Delta]) - 4(1-\cos[k_y\,Æ\,\Delta])\}\,\hat{u} \quad (A.8)$$

and

$$D_4 = D_2^2. \quad (A.9)$$

The semi-discrete equation is then

$$\Delta t \frac{d\hat{u}}{dt} = \lambda\,[-R + \epsilon_2 D_2 - \epsilon_4 D_4]\quad, \quad (A.10)$$

where

$$\lambda = \frac{\Delta t \Delta}{A}\quad, \quad (1.11a)$$

$$\epsilon_2 = \mu_2 \Delta\quad, \quad (1.11b)$$

$$\epsilon_4 = \mu_4 \Delta^3\quad. \quad (1.11c)$$

This defines a residual operator

$$z = -R + \epsilon_2 D_2 - \epsilon_4 D_4. \quad (1.12)$$

The state vector is updated using the multi-stage scheme presented in Chapter 3:

$$\hat{u}^{(0)} = \hat{u}^n\quad, \quad (1.13a)$$

$$\hat{u}^{(1)} = \hat{u}^{(0)} - \alpha_1 \lambda z\quad, \quad (1.13b)$$

$$\hat{u}^{(2)} = \hat{u}^{(0)} - \alpha_2 \lambda z\quad, \quad (1.13c)$$

$$\hat{u}^{(3)} = \hat{u}^{(0)} - \alpha_3 \lambda z\quad, \quad (1.13d)$$

$$\hat{u}^{(4)} = \hat{u}^{(0)} - \alpha_4 \lambda z\quad, \quad (1.13e)$$

$$\hat{u}^{n+1} = \hat{u}^{(4)}\quad. \quad (1.13f)$$

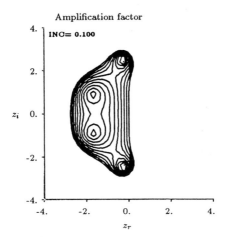

Figure A.1: Stability region for multi-stage

where

$$\alpha_1 = .25 \qquad \alpha_2 = .33 \qquad \alpha_3 = .50 \qquad \alpha_4 = 1.0 \quad . \tag{1.14}$$

The multi-stage as written above defines an amplification factor

$$G(\lambda z) = \frac{\hat{u}^{n+1}}{\hat{u}^n}. \tag{1.15}$$

Contours of $|G|$ in the z plane are shown in Figure A.1. The increment is 0.1 and the outermost contour is $|G| = 1$, the stability limit for the multi-stage. The contour intersects the imaginary axis at $z_i = 2\sqrt{2}$, which sets the CFL constraint, $\lambda < 2\sqrt{2}$.

Contours of $|G|$ in the wave-number plane for the undamped scheme on a cell of aspect ratio one are shown in Figure A.2. The contour increment is 0.1 for this and all wave-number plots. The amplification factor is one at the corners of the wave-number domain (necessary for consistency), at the center, and at $(\pi/3,\pi/3)$ and $(5\pi/3,5\pi/3)$. Figure A.3 shows contours of $|G|$ for a stretched cell, with $\mathcal{R} = 2$. On the stretched cell, the islands of maximum $|G|$ have shifted.

Figures A.4-A.7 show the effects of the fourth-difference damping operator. With a damping coefficient $\epsilon_4 = 0.001$, the locus of the wave number in the z plane (Figure A.4) falls within the stability boundary in Figure A.1. The amplification factor (Figure A.5) is everywhere less than one, reaching a value of 0.8 at $(\pi/3,\pi/3)$ and $(5\pi/3,5\pi/3)$. With a

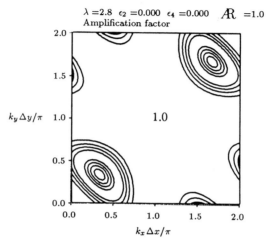

Figure A.2: Amplification factor contours — $\lambda = 1$, $\mathcal{R} = 1$, $\epsilon_2 = 0$, $\epsilon_4 = 0$

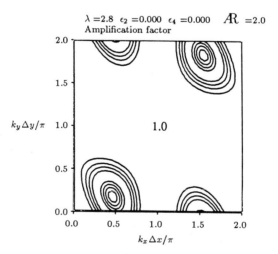

Figure A.3: Amplification factor contours — $\lambda = 1$, $\mathcal{R} = 2$, $\epsilon_2 = 0$, $\epsilon_4 = 0$

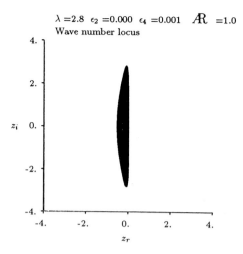

Figure A.4: Wave number locus — $\epsilon_2 = 0$, $\epsilon_4 = .001$, $\mathcal{R} = 1$

damping coefficient $\epsilon_4 = 0.005$, the locus of the wave number in the z plane (Figure A.6) exceeds the stability region, and the amplification factor (Figure A.7) exceeds one at $(\pi/2,\pi/2)$ and $(3\pi/2,3\pi/2)$.

Figures A.8-A.11 show the effects of the second-difference damping operator. With a damping coefficient $\epsilon_2 = 0.01$, the locus of the wave number in the z plane (Figure A.8) falls within the stability boundary in Figure A.1. The amplification factor (Figure A.9) is everywhere less than one, reaching a value of 0.7 at $(\pi/3,\pi/3)$ and $(5\pi/3,5\pi/3)$. With a damping coefficient $\epsilon_2 = 0.05$, the locus of the wave number in the z plane (Figure A.10) exceeds the stability region, and the amplification factor (Figure A.11) exceeds one at $(\pi/2,\pi/2)$ and $(3\pi/2,3\pi/2)$.

For the system of equations, the definition of λ is based on the eigenvalues of the Jacobian matrices. For the Euler equations, the eigenvalues of the Jacobians are $u_i n_i$, $u_i n_i + c$ and $u_i n_i - c$ where $u_i n_i$ is the flux through a face of the cell, and c is the local speed of sound [17]. This leads to the time-step constraint presented in Chapter 3.

198 APPENDIX A. STABILITY ANALYSIS

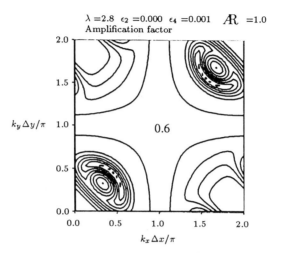

Figure A.5: Amplification factor contours — $\lambda = 1$, $\mathcal{R} = 1$, $\epsilon_2 = 0$, $\epsilon_4 = 0.001$

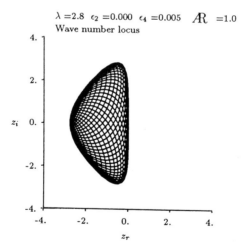

Figure A.6: Wave number locus — $\epsilon_2 = 0$, $\epsilon_4 = .005$, $\mathcal{R} = 1$

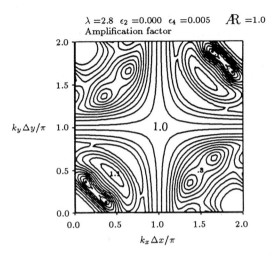

Figure A.7: Amplification factor contours — $\lambda = 1$, $Æ = 1$, $\epsilon_2 = 0$, $\epsilon_4 = 0.005$

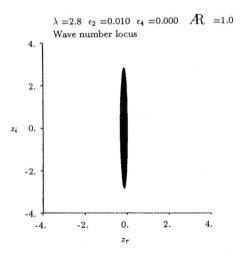

Figure A.8: Wave number locus — $\epsilon_2 = 0.01$, $\epsilon_4 = 0$, $Æ = 1$

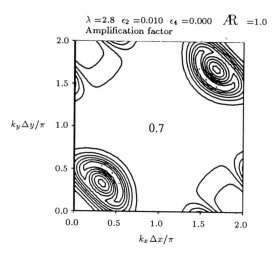

Figure A.9: Amplification factor contours — $\lambda = 1$, $\mathcal{R} = 1$, $\epsilon_2 = 0.01$, $\epsilon_4 = 0$

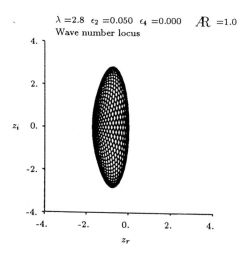

Figure A.10: Wave number locus — $\epsilon_2 = 0.05$, $\epsilon_4 = 0$, $\mathcal{R} = 1$

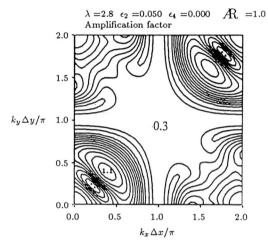

Figure A.11: Amplification factor contours — $\lambda = 1$, $Æ = 1$, $\epsilon_2 = 0.05$, $\epsilon_4 = 0$

Appendix B

Cross-flow streamline integration

The cross-flow streamline plots presented in this book are calculated by trajectory integrations of the cross-flow velocities $\bar{v} = v - \eta u$ and $\bar{w} = w - \zeta u$. The integration method used is of predictor-corrector type, stepping across a fraction of a cell at a time.

The i and j contravariant velocities are calculated at each node that is not on a grid boundary or embedded interface by the formulas

$$u_i = \frac{\bar{v}\frac{\partial i}{\partial \eta} + \bar{w}\frac{\partial i}{\partial \zeta}}{\sqrt{(1+\eta^2)\left(\frac{\partial i}{\partial \eta}\right)^2 + (1+\zeta^2)\left(\frac{\partial i}{\partial \zeta}\right)^2 + 2\eta\zeta\frac{\partial i}{\partial \eta}\frac{\partial i}{\partial \zeta}}}, \quad (2.1a)$$

$$u_j = \frac{\bar{v}\frac{\partial j}{\partial \eta} + \bar{w}\frac{\partial j}{\partial \zeta}}{\sqrt{(1+\eta^2)\left(\frac{\partial j}{\partial \eta}\right)^2 + (1+\zeta^2)\left(\frac{\partial j}{\partial \zeta}\right)^2 + 2\eta\zeta\frac{\partial j}{\partial \eta}\frac{\partial j}{\partial \zeta}}}. \quad (2.1b)$$

Values at boundary nodes and embedding interface nodes are obtained by a first-order extrapolation. The trajectory integration is carried out as follows:

1. Starting at a given point in a given cell, a cell-based coordinate system $(\hat{i}, \hat{j})$ is set up, where $-1 \leq \hat{i}, \hat{j} \leq 1$. The cell-based coordinates $(\hat{i}^0, \hat{j}^0)$ of the point are calculated by a bilinear interpolation in physical space;

2. The i and j contravariant velocities at the point are calculated by a bilinear interpolation in the computational space;

3. A predicted location for the next point on the trajectory is calculated by the formulas:

$$\hat{i}^p = \hat{i}^0 + \frac{u_i^0}{U}\Delta s, \quad (2.2a)$$

$$\hat{j}^p = \hat{j}^0 + \frac{u_j^0}{U}\Delta s, \quad (2.2b)$$

where $U = \max(|u_i^0|, |u_j^0|)$ is a reference velocity and Δs is an estimate of the fraction of the cell through which the trajectory will travel in one step (typically of order 10^{-1}).

4. The contravariant velocities at the predicted point u_i^p and u_j^p are calculated by a bilinear interpolation in the computational space.

5. A corrected location for the point is calculated by the formulas

$$\hat{i}^c = \hat{i}^0 + \frac{u_i^0 + u_i^p}{2U}\Delta s, \tag{2.3a}$$

$$\hat{j}^c = \hat{j}^0 + \frac{u_j^0 + u_j^p}{2U}\Delta s. \tag{2.3b}$$

6. If the trajectory integration has crossed a cell boundary, the new cell is found, and the cell-based coordinates $\hat{i}$ and $\hat{j}$ in the new cell are calculated.

7. The integration is ended if:

   - A boundary is crossed;
   - A cross-flow stagnation point is reached;
   - The trajectory reaches a limit cycle in the vortex.

Appendix C

Computer Code

This appendix contains the listing for LEVIS, the code used to calculate the symmetric flat plate cases presented in the book. The codes for the yaw and vortex flap cases differ only slightly from LEVIS, and will not be presented here. The modules of the program, the include file that contains dimensioning information and a sample input file are presented in the following pages.

Copyright ©1987 Massachusetts Institute of Technology

Permission to use, copy and modify this software and its documentation for internal purposes only and without fee is hereby granted provided that the above copyright notice and this permission appear on all copies of the code and supporting documentation. For any other use of this software, in original or modified form, including but not limited to, adaptation as the basis of a commerical software or hardware product, or distribution in whole or in part, specific prior permission and/or the appropriate license must be obtained from MIT. This software is provided "as is" without any warranties whatsoever, either express or implied, including but not limited to the implied warranties of merchantability and fitness for a particular purpose. This software is a research program, and MIT does not represent that it is free of errors or bugs or suitable for any particular task.

For further information, a listing, or a tape of LEVIS, please contact

M.I.T. Software Center

Technology Licensing Office

M.I.T. Room E32-300

28 Carleton Street

Cambridge, MA 02139.

A listing of LEVIS may be obtained for $20.00. A 9 track tape copy written in the tar format may be obtained for $200.00. Make checks payable to MIT and send to the above addresss. Please include your name and shipping address.

```
      PROGRAM LEVIS
C
C$$$$$$$$$$$$$$$$$$$$$$$$$$$$$$$$$$$$$$$$$$$$$$$$$$$$$$$$$$$$$$$$$$$$$
C$$$$$$$$$$$$$$$$$$$$$$$$$$$$$$$$$$$$$$$$$$$$$$$$$$$$$$$$$$$$$$$$$$$$$
C$$$$$                                                           $$$$$
C$$$$$    This is the program LEVIS, Leading Edge Vortex Inviscid $$$$$
C$$$$$    Solver. It solves the conical Euler equations.         $$$$$
C$$$$$                                                           $$$$$
C$$$$$$$$$$$$$$$$$$$$$$$$$$$$$$$$$$$$$$$$$$$$$$$$$$$$$$$$$$$$$$$$$$$$$
C$$$$$$$$$$$$$$$$$$$$$$$$$$$$$$$$$$$$$$$$$$$$$$$$$$$$$$$$$$$$$$$$$$$$$
C
      INCLUDE 'LEVIS.INC'
C
C.........  Read the input file.
C
      CALL READIN
C
C.........  Generate the grid and the pointers for a non-embedded grid.
C
      CALL GENPNT
      CALL GENGRD
C
C.........  Do embedding and set up boundary and iterface pointers.
C
      CALL GENEMB
      CALL GENBND
C
C.........  Calculate cell areas.
C
      CALL CALARS
C
C.........  Set the initial conditions.
C
      IF(RESTART) THEN
          CALL READRS
      ELSE
          CALL SETICS
      ENDIF
C
C.........  Start iteration loop.
C
      DO NITER=1,ITERS
          WRITE(6,*) 'ITERATION ',NITER,' PRESSURE',P(1)
          IF(NITER.NE.1) CALL CALRES
C
C.........  Calculate the time step.
C
          CALL CALDTS
```

```
C
C......... Save the state vector for the multi-stage.
C
          CALL SAVRES
C
C......... Multi-stage loop.
C
          DO KSTAGE=1,4
C
C......... Calculate the smoothing fluxes.
C
              CALL CALSMT
C
C......... Calculate the fluxes.
C
              CALL CALFLX
C
C......... Enforce the no-flux condition at the body.
C
              CALL FLUXBC
C
C......... Sum the fluxes.
C
              CALL SUMFLX
C
C......... Distribute the fluxes.
C
              CALL DISFLX
C
C......... Enforce the bc's at the outer boundary.
C
              CALL FARFBC
C
C......... Enforce the periodic boundary condition.
C
              CALL PERIBC
C
C......... Enforce the doubling condition at the body.
C
              CALL DOUBBC
C
C......... Update the state vectors.
C
              CALL UPDATE
C
C......... Calculate the pressures.
C
              CALL CALPRS
```

```
C
C.........  Next stage of the multi-stage
C
            ENDDO
C
C.........  Next iteration.
C
      ENDDO
C
C.........  Output solution.
C
      CALL OUTPUT
C
C
C
      END
```

```fortran
      SUBROUTINE CALARS
C
C$$$$$$$$$$$$$$$$$$$$$$$$$$$$$$$$$$$$$$$$$$$$$$$$$$$$$$$$$$$$$$$$$$$$
C$$$$$$$$$$$$$$$$$$$$$$$$$$$$$$$$$$$$$$$$$$$$$$$$$$$$$$$$$$$$$$$$$$$$
C$$$$$                                                          $$$$$
C$$$$$    This routine calculates the cell areas.               $$$$$
C$$$$$                                                          $$$$$
C$$$$$$$$$$$$$$$$$$$$$$$$$$$$$$$$$$$$$$$$$$$$$$$$$$$$$$$$$$$$$$$$$$$$
C$$$$$$$$$$$$$$$$$$$$$$$$$$$$$$$$$$$$$$$$$$$$$$$$$$$$$$$$$$$$$$$$$$$$
C
      INCLUDE 'LEVIS.INC'
C
C
C
      DO I=ICELBEG,ICELEND
C
          ISW=ICELNOD(1,I)
          ISE=ICELNOD(2,I)
          INE=ICELNOD(3,I)
          INW=ICELNOD(4,I)
C
          AREA(I)=-0.5*((X1(INE)-X1(ISW))*(X2(ISE)-X2(INW))-
     &                  (X2(INE)-X2(ISW))*(X1(ISE)-X1(INW)))
C
      ENDDO
C
C
C
      RETURN
      END
```

```
      SUBROUTINE CALDTS
C
C$$$$$$$$$$$$$$$$$$$$$$$$$$$$$$$$$$$$$$$$$$$$$$$$$$$$$$$$$$$$$$$$$$$$
C$$$$$$$$$$$$$$$$$$$$$$$$$$$$$$$$$$$$$$$$$$$$$$$$$$$$$$$$$$$$$$$$$$$$
C$$$$$                                                          $$$$$
C$$$$$    This subroutine calculates the maximum allowable time $$$$$
C$$$$$    step at each node.                                    $$$$$
C$$$$$                                                          $$$$$
C$$$$$$$$$$$$$$$$$$$$$$$$$$$$$$$$$$$$$$$$$$$$$$$$$$$$$$$$$$$$$$$$$$$$
C$$$$$$$$$$$$$$$$$$$$$$$$$$$$$$$$$$$$$$$$$$$$$$$$$$$$$$$$$$$$$$$$$$$$
C
      INCLUDE 'LEVIS.INC'
      REAL*8 DX1WE,DX2WE,DX1SN,DX2SN,
     &       DXSN,DXWE,DX1,DX2
C
C.........  Statement function for sound speed.
C
      C(I)=SQRT(GAMMA*P(I)/RHO(I))
C
C.........  Loop over the nodes.
C
      DO I=INODBEG,INODEND
C
C.........  Find the neighboring nodes.
C
         IS=INODNOD(1,I)
         IE=INODNOD(2,I)
         IN=INODNOD(3,I)
         IW=INODNOD(4,I)
C
C.........  Calculate the grid metrics.  Use centered differences.
C
         DX1WE=.5*(X1(IE)-X1(IW))
         DX2WE=.5*(X2(IE)-X2(IW))
         DX1SN=.5*(X1(IN)-X1(IS))
         DX2SN=.5*(X2(IN)-X2(IS))
C
C.........  Calculate the undirected side lengths.
C
         DXSN=ABS(DX1SN)+ABS(DX2SN)
         DXWE=ABS(DX1WE)+ABS(DX2WE)
C
         DX1=ABS(DX1WE)+ABS(DX1SN)
         DX2=ABS(DX2WE)+ABS(DX2SN)
C
C.........  Maximum eigenvalue.
C
         UX1=(RHOV(I)-X1(I)*RHOU(I))/SQRT(1+X1(I)**2)
```

```
              UX2=(RHOW(I)-X2(I)*RHOU(I))/SQRT(1+X2(I)**2)
              UDOTA=ABS(UX1)*DX1+ABS(UX2)*DX2
              CAMAG=C(I)*(DXSN+DXWE)
C
C.........  The time step.  2.8 is due to multi-stage coefficients.
C
              DT(I)=CFL*2.8/(UDOTA+CAMAG)
C
          ENDDO
C
C.........  Extrapolate interface time steps.  Suppress dependency check.
C
CVD$  NODEPCHK
          DO I=1,NCELINT
              INODE=ICELNODI(I)
              ITYPE=ICELFACI(2,I)
              IF(ITYPE.EQ.1) THEN
                  IW=INODNOD(4,INODE)
                  IE=INODNOD(2,INODE)
                  IS=INODNOD(1,INODE)
                  DT(INODE)=DT(IS)
                  DT(IW)=DT(INODE)
                  DT(IE)=DT(INODE)
              ELSEIF(ITYPE.EQ.2) THEN
                  IS=INODNOD(1,INODE)
                  IN=INODNOD(3,INODE)
                  IE=INODNOD(2,INODE)
                  DT(INODE)=DT(IE)
                  DT(IS)=DT(INODE)
                  DT(IN)=DT(INODE)
              ELSEIF(ITYPE.EQ.3) THEN
                  IE=INODNOD(2,INODE)
                  IW=INODNOD(4,INODE)
                  IN=INODNOD(3,INODE)
                  DT(INODE)=DT(IN)
                  DT(IE)=DT(INODE)
                  DT(IW)=DT(INODE)
              ELSEIF(ITYPE.EQ.4) THEN
                  IN=INODNOD(3,INODE)
                  IS=INODNOD(1,INODE)
                  IW=INODNOD(4,INODE)
                  DT(INODE)=DT(IW)
                  DT(IN)=DT(INODE)
                  DT(IS)=DT(INODE)
              ENDIF
          ENDDO
C
C.........  Extrapolate the boundary time steps.  Suppress dependency check.
```

```
C
CVD$  NODEPCHK
      DO INODE=1,NNODFF
          I=INODFF(INODE)
          IS=INODNOD(1,I)
          DT(I)=DT(IS)
      ENDDO
C
CVD$  NODEPCHK
      DO INODE=1,NNODPW
          I=INODPW(INODE)
          IE=INODNOD(2,I)
          DT(I)=DT(IE)
      ENDDO
C
CVD$  NODEPCHK
      DO INODE=1,NNODPE
          I=INODPE(INODE)
          IW=INODNOD(4,I)
          DT(I)=DT(IW)
      ENDDO
C
CVD$  NODEPCHK
      DO INODE=1,NNODWL
          I=INODWL(INODE)
          IN=INODNOD(3,I)
          DT(I)=DT(IN)
      ENDDO
C
C
C
      RETURN
      END
```

```fortran
      SUBROUTINE CALFLX
C
C$$$$$$$$$$$$$$$$$$$$$$$$$$$$$$$$$$$$$$$$$$$$$$$$$$$$$$$$$$$$$$$$$$$$
C$$$$$$$$$$$$$$$$$$$$$$$$$$$$$$$$$$$$$$$$$$$$$$$$$$$$$$$$$$$$$$$$$$$$
C$$$$$                                                          $$$$$
C$$$$$    This subroutine calculates the fluxes on each face.   $$$$$
C$$$$$                                                          $$$$$
C$$$$$$$$$$$$$$$$$$$$$$$$$$$$$$$$$$$$$$$$$$$$$$$$$$$$$$$$$$$$$$$$$$$$
C$$$$$$$$$$$$$$$$$$$$$$$$$$$$$$$$$$$$$$$$$$$$$$$$$$$$$$$$$$$$$$$$$$$$
C
      INCLUDE 'LEVIS.INC'
      REAL UB(NNODESMX),
     &     F1(NNODESMX),
     &     F2(NNODESMX),
     &     F3(NNODESMX),
     &     F4(NNODESMX),
     &     F5(NNODESMX),
     &     VB(NNODESMX),
     &     G1(NNODESMX),
     &     G2(NNODESMX),
     &     G3(NNODESMX),
     &     G4(NNODESMX),
     &     G5(NNODESMX)
C
C.........  Calculate the flux vectors.
C
      DO I=INODBEG,INODEND
         UB(I)=(RHOV(I)-X1(I)*RHOU(I))/RHO(I)
         F1(I)=RHO (I)*UB(I)
         F2(I)=RHOU(I)*UB(I)-X1(I)*P(I)
         F3(I)=RHOV(I)*UB(I)+        P(I)
         F4(I)=RHOW(I)*UB(I)
         F5(I)=RHOE(I)*UB(I)+UB(I)*P(I)
C
         VB(I)=(RHOW(I)-X2(I)*RHOU(I))/RHO(I)
         G1(I)=RHO (I)*VB(I)
         G2(I)=RHOU(I)*VB(I)-X2(I)*P(I)
         G3(I)=RHOV(I)*VB(I)
         G4(I)=RHOW(I)*VB(I)+        P(I)
         G5(I)=RHOE(I)*VB(I)+VB(I)*P(I)
      ENDDO
C
C.........  Visit the north-soutn faces and calculate the fluxes.
C
      DO IFACE=IFA1BEG,IFA1END
C
C.........  Find the two nodes that define the face.
C
```

```
              INODE1=IFA1NOD(1,IFACE)
              INODE2=IFA1NOD(2,IFACE)
C
C.........  Calculate the face normal.
C
              DX1=X1(INODE2)-X1(INODE1)
              DX2=X2(INODE2)-X2(INODE1)
C
C.........  Trapezoidal integration for the fluxes.
C
              FLUX1(1,IFACE)=.5*( (F1(INODE1)+F1(INODE2))*DX2 -
     &                            (G1(INODE1)+G1(INODE2))*DX1   )
              FLUX2(1,IFACE)=.5*( (F2(INODE1)+F2(INODE2))*DX2 -
     &                            (G2(INODE1)+G2(INODE2))*DX1   )
              FLUX3(1,IFACE)=.5*( (F3(INODE1)+F3(INODE2))*DX2 -
     &                            (G3(INODE1)+G3(INODE2))*DX1   )
              FLUX4(1,IFACE)=.5*( (F4(INODE1)+F4(INODE2))*DX2 -
     &                            (G4(INODE1)+G4(INODE2))*DX1   )
              FLUX5(1,IFACE)=.5*( (F5(INODE1)+F5(INODE2))*DX2 -
     &                            (G5(INODE1)+G5(INODE2))*DX1   )
C
       ENDDO
C
C.........  Visit the east-west faces and calculate the fluxes.
C
       DO IFACE=IFA2BEG,IFA2END
C
C.........  Find the two nodes that define the face.
C
              INODE1=IFA2NOD(1,IFACE)
              INODE2=IFA2NOD(2,IFACE)
C
C.........  Calculate the face normal.
C
              DX1=X1(INODE2)-X1(INODE1)
              DX2=X2(INODE2)-X2(INODE1)
C
C.........  Trapezoidal integration for the fluxes.
C
              FLUX1(2,IFACE)=.5*( (F1(INODE1)+F1(INODE2))*DX2 -
     &                            (G1(INODE1)+G1(INODE2))*DX1   )
              FLUX2(2,IFACE)=.5*( (F2(INODE1)+F2(INODE2))*DX2 -
     &                            (G2(INODE1)+G2(INODE2))*DX1   )
              FLUX3(2,IFACE)=.5*( (F3(INODE1)+F3(INODE2))*DX2 -
     &                            (G3(INODE1)+G3(INODE2))*DX1   )
              FLUX4(2,IFACE)=.5*( (F4(INODE1)+F4(INODE2))*DX2 -
     &                            (G4(INODE1)+G4(INODE2))*DX1   )
              FLUX5(2,IFACE)=.5*( (F5(INODE1)+F5(INODE2))*DX2 -
```

```
     &              (G5(INODE1)+G5(INODE2))*DX1   )
C
      ENDDO
C
C
C
      RETURN
      END
```

```
      SUBROUTINE CALPRS
C
C$$$$$$$$$$$$$$$$$$$$$$$$$$$$$$$$$$$$$$$$$$$$$$$$$$$$$$$$$$$$$$$$$$$$$
C$$$$$$$$$$$$$$$$$$$$$$$$$$$$$$$$$$$$$$$$$$$$$$$$$$$$$$$$$$$$$$$$$$$$$
C$$$$$                                                           $$$$$
C$$$$$   This subroutine calculates the pressure.  If the pressure $$$$$
C$$$$$   is zero, the state vectors are "un-updated" to avoid    $$$$$
C$$$$$   bad transients.                                         $$$$$
C$$$$$                                                           $$$$$
C$$$$$$$$$$$$$$$$$$$$$$$$$$$$$$$$$$$$$$$$$$$$$$$$$$$$$$$$$$$$$$$$$$$$$
C$$$$$$$$$$$$$$$$$$$$$$$$$$$$$$$$$$$$$$$$$$$$$$$$$$$$$$$$$$$$$$$$$$$$$
C
      INCLUDE 'LEVIS.INC'
C
C.......... Loop over the nodes.
C
      EPS=1.0E-6
C
      DO INODE=INODBEG,INODEND
         P(INODE)=(GAMMA-1)*(RHOE(INODE)-
     &           .5*(RHOU(INODE)**2+RHOV(INODE)**2+
     &               RHOW(INODE)**2)/RHO(INODE))
         IF(P(INODE).LE.EPS.OR.RHO(INODE).LE.EPS) THEN
            RHO (INODE)=RHOO (INODE)
            RHOU(INODE)=RHOUO(INODE)
            RHOV(INODE)=RHOVO(INODE)
            RHOW(INODE)=RHOWO(INODE)
            RHOE(INODE)=RHOEO(INODE)
            P(INODE)=(GAMMA-1)*(RHOE(INODE)-
     &              .5*(RHOU(INODE)**2+RHOV(INODE)**2+
     &                  RHOW(INODE)**2)/RHO(INODE))
         ENDIF
      ENDDO
C
C
C
      RETURN
      END
```

```
      SUBROUTINE CALRES
C
C$$$$$$$$$$$$$$$$$$$$$$$$$$$$$$$$$$$$$$$$$$$$$$$$$$$$$$$$$$$$$$$$$$$$
C$$$$$$$$$$$$$$$$$$$$$$$$$$$$$$$$$$$$$$$$$$$$$$$$$$$$$$$$$$$$$$$$$$$$
C$$$$$                                                          $$$$$
C$$$$$    This subroutine calculates the residuals at each iter- $$$$$
C$$$$$    ation and writes them to a convergence history file.  $$$$$
C$$$$$                                                          $$$$$
C$$$$$$$$$$$$$$$$$$$$$$$$$$$$$$$$$$$$$$$$$$$$$$$$$$$$$$$$$$$$$$$$$$$$
C$$$$$$$$$$$$$$$$$$$$$$$$$$$$$$$$$$$$$$$$$$$$$$$$$$$$$$$$$$$$$$$$$$$$
C
      INCLUDE 'LEVIS.INC'
C
C......... Initialize the L-2 and L-infinity residuals.
C
      RESL2=0
      RESLI=0
C
C......... Loop over the nodes.
C
      DO INODE=INODBEG,INODEND
C
C......... L-2 norm.
C
          RESL2=RESL2+(RHO (INODE)-RHOO (INODE))**2
C
C......... L-infinity norm and node at which it occurs.
C
          RESTM=   ABS(RHO (INODE)-RHOO (INODE))/RHO(INODE)
          IF(RESTM.GT.RESLI) THEN
              RESLI=RESTM
              IMXRS=INODE
          ENDIF
C
      ENDDO
C
C......... Normalize the L-2 norm.
C
      NNODES=INODEND-INODBEG
      RESL2=SQRT(RESL2/NNODES)
C
C......... Write the norms to the output device and the residual file.
C
      WRITE(6,*) 'L2 RES=',RESL2,'. LINF RES=',RESLI,' AT',IMXRS
      OPEN(UNIT=2,FILE='RESID.DAT',ACCESS='APPEND',STATUS='UNKNOWN')
      WRITE(2,*) RESL2,RESLI,IMXRS
      CLOSE(UNIT=2)
C
```

```
C
C
      RETURN
      END
```

```
      SUBROUTINE CALSMT
C
C$$$$$$$$$$$$$$$$$$$$$$$$$$$$$$$$$$$$$$$$$$$$$$$$$$$$$$$$$$$$$$$$$$$$$
C$$$$$$$$$$$$$$$$$$$$$$$$$$$$$$$$$$$$$$$$$$$$$$$$$$$$$$$$$$$$$$$$$$$$$
C$$$$$                                                           $$$$$
C$$$$$   This subroutine calculates the damping.                 $$$$$
C$$$$$                                                           $$$$$
C$$$$$$$$$$$$$$$$$$$$$$$$$$$$$$$$$$$$$$$$$$$$$$$$$$$$$$$$$$$$$$$$$$$$$
C$$$$$$$$$$$$$$$$$$$$$$$$$$$$$$$$$$$$$$$$$$$$$$$$$$$$$$$$$$$$$$$$$$$$$
C
      INCLUDE 'LEVIS.INC'
      REAL RHOH (NNODESMX),
     &     TEMP1(NNODESMX),
     &     TEMP2(NNODESMX)
C
C.........  Calculate rhoh for energy equation smoothing.
C
      DO INODE=INODBEG,INODEND
         RHOH(INODE)=RHOE(INODE)+P(INODE)
      ENDDO
C
C.........  Calculate unweighted second difference of pressure.
C
      DO INODE=INODBEG,INODEND
         TEMP2(INODE)=P(INODE)
      ENDDO
      CALL SECDNW(TEMP2,TEMP1)
      DO INODE=INODBEG,INODEND
         DP(INODE)=ABS(TEMP1(INODE))/P(INODE)
      ENDDO
C
C.........  Smooth the pressure switch.
C
      CALL SECDNW(DP,TEMP1)
      DO INODE=INODBEG,INODEND
         DP(INODE)=DP(INODE)+0.05*TEMP1(INODE)
      ENDDO
C
C.........  Extrapolate interface switches.  Suppress dependency check.
C
CVD$ NODEPCHK
      DO I=1,NCELINT
         INODE=ICELNODI(I)
         ITYPE=ICELFACI(2,I)
         IF(ITYPE.EQ.1) THEN
            IW=INODNOD(4,INODE)
            IE=INODNOD(2,INODE)
            IS=INODNOD(1,INODE)
```

```
                    DP(INODE)=DP(IS)
                    DP(IW)=DP(INODE)
                    DP(IE)=DP(INODE)
                ELSEIF(ITYPE.EQ.2) THEN
                    IS=INODNOD(1,INODE)
                    IN=INODNOD(3,INODE)
                    IE=INODNOD(2,INODE)
                    DP(INODE)=DP(IE)
                    DP(IS)=DP(INODE)
                    DP(IN)=DP(INODE)
                ELSEIF(ITYPE.EQ.3) THEN
                    IE=INODNOD(2,INODE)
                    IW=INODNOD(4,INODE)
                    IN=INODNOD(3,INODE)
                    DP(INODE)=DP(IN)
                    DP(IE)=DP(INODE)
                    DP(IW)=DP(INODE)
                ELSEIF(ITYPE.EQ.4) THEN
                    IN=INODNOD(3,INODE)
                    IS=INODNOD(1,INODE)
                    IW=INODNOD(4,INODE)
                    DP(INODE)=DP(IW)
                    DP(IN)=DP(INODE)
                    DP(IS)=DP(INODE)
                ENDIF
            ENDDO
C
C.........  Extrapolate the boundary switches.  Suppress dependency check.
C
CVD$    NODEPCHK
            DO INODE=1,NNODFF
                I=INODFF(INODE)
                IS=INODNOD(1,I)
                DP(I)=DP(IS)
            ENDDO
C
CVD$    NODEPCHK
            DO INODE=1,NNODPW
                I=INODPW(INODE)
                IE=INODNOD(2,I)
                DP(I)=DP(IE)
            ENDDO
C
CVD$    NODEPCHK
            DO INODE=1,NNODPE
                I=INODPE(INODE)
                IW=INODNOD(4,I)
                DP(I)=DP(IW)
```

```
            ENDDO
      C
      CVD$  NODEPCHK
            DO INODE=1,NNODWL
                I=INODWL(INODE)
                IN=INODNOD(3,I)
                DP(I)=DP(IN)
            ENDDO
      C
      C.........  Find the maximum pressure switch.
      C
            DPMAX=1.0E-4
            DO INODE=INODBEG,INODEND
                DPMAX=MAX(DPMAX,DP(INODE))
            ENDDO
      C
      C.........  Normalize the pressures switches by the maximum value.
      C
            DO INODE=INODBEG,INODEND
                DP(INODE)=DP(INODE)/DPMAX
            ENDDO
      C
      C.........  Override the appropriate switches, setting them to one.
      C
            DO I=1,NNODOV
                INODE=INODOV(I)
                DP(INODE)=1.0
            ENDDO
      C
      C.........  Calculate the pressure-weighted second-difference damping.
      C
            CALL SECDIF(RHO ,DP,TEMP1)
            DO INODE=INODBEG,INODEND
                DAMP1(INODE)=EPS2*TEMP1(INODE)
            ENDDO
      C
            CALL SECDIF(RHOU,DP,TEMP1)
            DO INODE=INODBEG,INODEND
                DAMP2(INODE)=EPS2*TEMP1(INODE)
            ENDDO
      C
            CALL SECDIF(RHOV,DP,TEMP1)
            DO INODE=INODBEG,INODEND
                DAMP3(INODE)=EPS2*TEMP1(INODE)
            ENDDO
      C
            CALL SECDIF(RHOW,DP,TEMP1)
            DO INODE=INODBEG,INODEND
```

```
              DAMP4(INODE)=EPS2*TEMP1(INODE)
          ENDDO
C
          CALL SECDIF(RHOH,DP,TEMP1)
          DO INODE=INODBEG,INODEND
              DAMP5(INODE)=EPS2*TEMP1(INODE)
          ENDDO
C
C.......... Calculate the unweighted fourth difference and subtract.
C
          CALL SECDNW(RHO  ,TEMP1)
          CALL SECDNW(TEMP1,TEMP2)
          DO INODE=INODBEG,INODEND
              DAMP1(INODE)=DAMP1(INODE)-EPS4*TEMP2(INODE)
          ENDDO
C
          CALL SECDNW(RHOU ,TEMP1)
          CALL SECDNW(TEMP1,TEMP2)
          DO INODE=INODBEG,INODEND
              DAMP2(INODE)=DAMP2(INODE)-EPS4*TEMP2(INODE)
          ENDDO
C
          CALL SECDNW(RHOV ,TEMP1)
          CALL SECDNW(TEMP1,TEMP2)
          DO INODE=INODBEG,INODEND
              DAMP3(INODE)=DAMP3(INODE)-EPS4*TEMP2(INODE)
          ENDDO
C
          CALL SECDNW(RHOW ,TEMP1)
          CALL SECDNW(TEMP1,TEMP2)
          DO INODE=INODBEG,INODEND
              DAMP4(INODE)=DAMP4(INODE)-EPS4*TEMP2(INODE)
          ENDDO
C
          CALL SECDNW(RHOH ,TEMP1)
          CALL SECDNW(TEMP1,TEMP2)
          DO INODE=INODBEG,INODEND
              DAMP5(INODE)=DAMP5(INODE)-EPS4*TEMP2(INODE)
          ENDDO
C
C
C
          RETURN
          END
```

```
      SUBROUTINE CELCEL(ITYPE,I,IOUT)
C
C$$$$$$$$$$$$$$$$$$$$$$$$$$$$$$$$$$$$$$$$$$$$$$$$$$$$$$$$$$$$$$$$$$$$$
C$$$$$$$$$$$$$$$$$$$$$$$$$$$$$$$$$$$$$$$$$$$$$$$$$$$$$$$$$$$$$$$$$$$$$
C$$$$$                                                           $$$$$
C$$$$$    This subroutine finds a neighboring cell.              $$$$$
C$$$$$                                                           $$$$$
C$$$$$$$$$$$$$$$$$$$$$$$$$$$$$$$$$$$$$$$$$$$$$$$$$$$$$$$$$$$$$$$$$$$$$
C$$$$$$$$$$$$$$$$$$$$$$$$$$$$$$$$$$$$$$$$$$$$$$$$$$$$$$$$$$$$$$$$$$$$$
C
      INCLUDE 'LEVIS.INC'
C
C.......... Look for cell to south
C
      IF(ITYPE.EQ.1) THEN
C
          IFACE=ICELFAC(1,I)
          DO J=ICELBEG,ICELEND
             IF(ICELFAC(3,J).EQ.IFACE) THEN
                IOUT=J
                GO TO 1
             ENDIF
          ENDDO
  1       CONTINUE
C
C.......... Look for cell to east
C
      ELSEIF(ITYPE.EQ.2) THEN
C
          IFACE=ICELFAC(2,I)
          DO J=ICELBEG,ICELEND
             IF(ICELFAC(4,J).EQ.IFACE) THEN
                IOUT=J
                GO TO 2
             ENDIF
          ENDDO
  2       CONTINUE
C
C.......... Look for cell to north
C
      ELSEIF(ITYPE.EQ.3) THEN
C
          IFACE=ICELFAC(3,I)
          DO J=ICELBEG,ICELEND
             IF(ICELFAC(1,J).EQ.IFACE) THEN
                IOUT=J
                GO TO 3
             ENDIF
```

```
              ENDDO
   3          CONTINUE
C
C.......... Look for cell to west
C
              ELSEIF(ITYPE.EQ.4) THEN
C
              IFACE=ICELFAC(4,I)
              DO J=ICELBEG,ICELEND
                 IF(ICELFAC(2,J).EQ.IFACE) THEN
                    IOUT=J
                    GO TO 4
                 ENDIF
              ENDDO
   4          CONTINUE
C
        ENDIF
C
C
C       RETURN
        END
```

```
      SUBROUTINE DISFLX
C
C$$$$$$$$$$$$$$$$$$$$$$$$$$$$$$$$$$$$$$$$$$$$$$$$$$$$$$$$$$$$$$$$$$$$
C$$$$$$$$$$$$$$$$$$$$$$$$$$$$$$$$$$$$$$$$$$$$$$$$$$$$$$$$$$$$$$$$$$$$
C$$$$$                                                          $$$$$
C$$$$$   This subroutine distributes the fluxes to the nodes.   $$$$$
C$$$$$   The distribution to the four nodes is split into four  $$$$$
C$$$$$   loops to prohibit dependencies.                        $$$$$
C$$$$$                                                          $$$$$
C$$$$$$$$$$$$$$$$$$$$$$$$$$$$$$$$$$$$$$$$$$$$$$$$$$$$$$$$$$$$$$$$$$$$
C$$$$$$$$$$$$$$$$$$$$$$$$$$$$$$$$$$$$$$$$$$$$$$$$$$$$$$$$$$$$$$$$$$$$
C
      INCLUDE 'LEVIS.INC'
C
C.........  Visit the nodes.  Zero the changes.
C
      DO INODE=INODBEG,INODEND
         DR (INODE)=0
         DRU(INODE)=0
         DRV(INODE)=0
         DRW(INODE)=0
         DRE(INODE)=0
      ENDDO
C
C.........  Visit the cells.  Distribute the changes to the sw nodes.
C
CVD$  NODEPCHK
      DO ICELL=ICELBEG,ICELEND
C
         ISW=ICELNOD(1,ICELL)
C
C.........  Distribute the changes.
C
         DR (ISW)=DR (ISW)+.25*SFLUX1(ICELL)
         DRU(ISW)=DRU(ISW)+.25*SFLUX2(ICELL)
         DRV(ISW)=DRV(ISW)+.25*SFLUX3(ICELL)
         DRW(ISW)=DRW(ISW)+.25*SFLUX4(ICELL)
         DRE(ISW)=DRE(ISW)+.25*SFLUX5(ICELL)
C
      ENDDO
C
C.........  Visit the cells.  Distribute the changes to the se nodes.
C
CVD$  NODEPCHK
      DO ICELL=ICELBEG,ICELEND
C
         ISE=ICELNOD(2,ICELL)
C
```

C......... Distribute the changes.
C
            DR (ISE)=DR (ISE)+.25*SFLUX1(ICELL)
            DRU(ISE)=DRU(ISE)+.25*SFLUX2(ICELL)
            DRV(ISE)=DRV(ISE)+.25*SFLUX3(ICELL)
            DRW(ISE)=DRW(ISE)+.25*SFLUX4(ICELL)
            DRE(ISE)=DRE(ISE)+.25*SFLUX5(ICELL)
C
      ENDDO
C
C......... Visit the cells. Distribute the changes to the ne nodes.
C
CVD$ NODEPCHK
      DO ICELL=ICELBEG,ICELEND
C
            INE=ICELNOD(3,ICELL)
C
C......... Distribute the changes.
C
            DR (INE)=DR (INE)+.25*SFLUX1(ICELL)
            DRU(INE)=DRU(INE)+.25*SFLUX2(ICELL)
            DRV(INE)=DRV(INE)+.25*SFLUX3(ICELL)
            DRW(INE)=DRW(INE)+.25*SFLUX4(ICELL)
            DRE(INE)=DRE(INE)+.25*SFLUX5(ICELL)
C
      ENDDO
C
C......... Visit the cells. Distribute the changes to the nw nodes.
C
CVD$ NODEPCHK
      DO ICELL=ICELBEG,ICELEND
C
            INW=ICELNOD(4,ICELL)
C
C......... Distribute the changes.
C
            DR (INW)=DR (INW)+.25*SFLUX1(ICELL)
            DRU(INW)=DRU(INW)+.25*SFLUX2(ICELL)
            DRV(INW)=DRV(INW)+.25*SFLUX3(ICELL)
            DRW(INW)=DRW(INW)+.25*SFLUX4(ICELL)
            DRE(INW)=DRE(INW)+.25*SFLUX5(ICELL)
C
      ENDDO
C
C......... Interface cells. Distribute the changes to the hanging nodes.
C
      FAC1=SQRT(2.0)/(4+2*SQRT(2.0))
      FAC2=.5/(2+SQRT(2.0))-.25

```
      DO I=1,NCELINT
         ICELL=ICELINT(I)
         INODE=ICELNODI(I)
         ITYPE=ICELFACI(2,I)
         DR (INODE)=DR (INODE)+FAC1*SFLUX1(ICELL)
         DRU(INODE)=DRU(INODE)+FAC1*SFLUX2(ICELL)
         DRV(INODE)=DRV(INODE)+FAC1*SFLUX3(ICELL)
         DRW(INODE)=DRW(INODE)+FAC1*SFLUX4(ICELL)
         DRE(INODE)=DRE(INODE)+FAC1*SFLUX5(ICELL)
         IF(ITYPE.EQ.1.OR.ITYPE.EQ.3) THEN
            IW=INODNOD(4,INODE)
            IE=INODNOD(2,INODE)
            DR (IW)=DR (IW)+FAC2*SFLUX1(ICELL)
            DRU(IW)=DRU(IW)+FAC2*SFLUX2(ICELL)
            DRV(IW)=DRV(IW)+FAC2*SFLUX3(ICELL)
            DRW(IW)=DRW(IW)+FAC2*SFLUX4(ICELL)
            DRE(IW)=DRE(IW)+FAC2*SFLUX5(ICELL)
            DR (IE)=DR (IE)+FAC2*SFLUX1(ICELL)
            DRU(IE)=DRU(IE)+FAC2*SFLUX2(ICELL)
            DRV(IE)=DRV(IE)+FAC2*SFLUX3(ICELL)
            DRW(IE)=DRW(IE)+FAC2*SFLUX4(ICELL)
            DRE(IE)=DRE(IE)+FAC2*SFLUX5(ICELL)
         ELSEIF(ITYPE.EQ.2.OR.ITYPE.EQ.4) THEN
            IS=INODNOD(1,INODE)
            IN=INODNOD(3,INODE)
            DR (IS)=DR (IS)+FAC2*SFLUX1(ICELL)
            DRU(IS)=DRU(IS)+FAC2*SFLUX2(ICELL)
            DRV(IS)=DRV(IS)+FAC2*SFLUX3(ICELL)
            DRW(IS)=DRW(IS)+FAC2*SFLUX4(ICELL)
            DRE(IS)=DRE(IS)+FAC2*SFLUX5(ICELL)
            DR (IN)=DR (IN)+FAC2*SFLUX1(ICELL)
            DRU(IN)=DRU(IN)+FAC2*SFLUX2(ICELL)
            DRV(IN)=DRV(IN)+FAC2*SFLUX3(ICELL)
            DRW(IN)=DRW(IN)+FAC2*SFLUX4(ICELL)
            DRE(IN)=DRE(IN)+FAC2*SFLUX5(ICELL)
         ENDIF
      ENDDO
C
C
C
      RETURN
      END
```

```
      SUBROUTINE DIVCEL(ILEM,ICELL,IFLAG,
     &                       INODEND1,ICELEND1,IFA1END1,IFA2END1)
C
C$$$$$$$$$$$$$$$$$$$$$$$$$$$$$$$$$$$$$$$$$$$$$$$$$$$$$$$$$$$$$$$$$$$
C$$$$$$$$$$$$$$$$$$$$$$$$$$$$$$$$$$$$$$$$$$$$$$$$$$$$$$$$$$$$$$$$$$$
C$$$$$                                                         $$$$$
C$$$$$    This subroutine divides a cell, adding nodes and faces.  $$$$$
C$$$$$    Temporary pointers (ending in 1) are used to get node,   $$$$$
C$$$$$    cell and face indices.  IFLAG is: 1 for a cell to be     $$$$$
C$$$$$    divided; 2 for a child cell that has been introduced; 3  $$$$$
C$$$$$    for a parent cell that has been divided.                 $$$$$
C$$$$$                                                         $$$$$
C$$$$$$$$$$$$$$$$$$$$$$$$$$$$$$$$$$$$$$$$$$$$$$$$$$$$$$$$$$$$$$$$$$$
C$$$$$$$$$$$$$$$$$$$$$$$$$$$$$$$$$$$$$$$$$$$$$$$$$$$$$$$$$$$$$$$$$$$
C
      INCLUDE 'LEVIS.INC'
      INTEGER IFLAG(NNODESMX)
      IDUMMY=NNODESMX
C
C......... Find cell to south.
C
      ICELLS=IDUMMY
      IFACE=ICELFAC(1,ICELL)
      DO I=ICELBEG,ICELEND
         IF(ICELFAC(3,I).EQ.IFACE) THEN
            ICELLS=I
            GO TO 1
         ENDIF
      ENDDO
  1   CONTINUE
C
C......... Find cell to north.
C
      ICELLN=IDUMMY
      IFACE=ICELFAC(3,ICELL)
      DO I=ICELBEG,ICELEND
         IF(ICELFAC(1,I).EQ.IFACE) THEN
            ICELLN=I
            GO TO 2
         ENDIF
      ENDDO
  2   CONTINUE
C
C......... Find cell to west.
C
      ICELLW=IDUMMY
      IFACE=ICELFAC(4,ICELL)
      DO I=ICELBEG,ICELEND
```

```
              IF(ICELFAC(2,I).EQ.IFACE) THEN
                  ICELLW=I
                  GO TO 3
              ENDIF
          ENDDO
 3        CONTINUE
C
C......... Find cell to east.
C
          ICELLE=IDUMMY
          IFACE=ICELFAC(2,ICELL)
          DO I=ICELBEG,ICELEND
              IF(ICELFAC(4,I).EQ.IFACE) THEN
                  ICELLE=I
                  GO TO 4
              ENDIF
          ENDDO
 4        CONTINUE
C
C......... Divide icell into four cells.
C
          ICELL1=ICELL
          ICELL2=ICELEND1+1
          ICELL3=ICELEND1+2
          ICELL4=ICELEND1+3
C
          ICELEND1=ICELEND1+3
C
C......... Cell-to-node pointer.
C
          ICELNOD(1,ICELL1)=ICELNOD(1,ICELL1)
          ICELNOD(2,ICELL2)=ICELNOD(2,ICELL1)
          ICELNOD(3,ICELL3)=ICELNOD(3,ICELL1)
          ICELNOD(4,ICELL4)=ICELNOD(4,ICELL1)
C
C......... Check for type of cell and add nodes.
C
          IF(IFLAG(ICELLS).EQ.0.AND.IFLAG(ICELLW).EQ.0) THEN
C
              ICELNOD(2,ICELL1)=INODEND1+1
              ICELNOD(3,ICELL1)=INODEND1+2
              ICELNOD(4,ICELL1)=INODEND1+3
              ICELNOD(1,ICELL2)=INODEND1+1
              ICELNOD(3,ICELL2)=INODEND1+4
              ICELNOD(4,ICELL2)=INODEND1+2
              ICELNOD(1,ICELL3)=INODEND1+2
              ICELNOD(2,ICELL3)=INODEND1+4
              ICELNOD(4,ICELL3)=INODEND1+5
```

```
              ICELNOD(1,ICELL4)=INODEND1+3
              ICELNOD(2,ICELL4)=INODEND1+2
              ICELNOD(3,ICELL4)=INODEND1+5
C
              INODEND1=INODEND1+5
C
         ELSEIF(IFLAG(ICELLS).EQ.0) THEN
C
              ICELNOD(2,ICELL1)=INODEND1+1
              ICELNOD(3,ICELL1)=INODEND1+2
              ICELNOD(1,ICELL2)=INODEND1+1
              ICELNOD(3,ICELL2)=INODEND1+3
              ICELNOD(4,ICELL2)=INODEND1+2
              ICELNOD(1,ICELL3)=INODEND1+2
              ICELNOD(2,ICELL3)=INODEND1+3
              ICELNOD(4,ICELL3)=INODEND1+4
              ICELNOD(2,ICELL4)=INODEND1+2
              ICELNOD(3,ICELL4)=INODEND1+4
C
              INODEND1=INODEND1+4
C
C.......... Get node from cell to west.
C
              IF(IFLAG(ICELLW).EQ.2) THEN
                  ICELNOD(4,ICELL1)=ICELNOD(3,ICELLW)
                  ICELNOD(1,ICELL4)=ICELNOD(4,ICELL1)
              ENDIF
C
         ELSEIF(IFLAG(ICELLW).EQ.0) THEN
C
              ICELNOD(3,ICELL1)=INODEND1+1
              ICELNOD(4,ICELL1)=INODEND1+2
              ICELNOD(3,ICELL2)=INODEND1+3
              ICELNOD(4,ICELL2)=INODEND1+1
              ICELNOD(1,ICELL3)=INODEND1+1
              ICELNOD(2,ICELL3)=INODEND1+3
              ICELNOD(4,ICELL3)=INODEND1+4
              ICELNOD(1,ICELL4)=INODEND1+2
              ICELNOD(2,ICELL4)=INODEND1+1
              ICELNOD(3,ICELL4)=INODEND1+4
C
              INODEND1=INODEND1+4
C
C.......... Get node from cell to south.
C
              IF(IFLAG(ICELLS).EQ.2) THEN
                  ICELNOD(2,ICELL1)=ICELNOD(3,ICELLS)
                  ICELNOD(1,ICELL2)=ICELNOD(2,ICELL1)
```

```
              ENDIF
C
        ELSE
C
              ICELNOD(3,ICELL1)=INODEND1+1
              ICELNOD(3,ICELL2)=INODEND1+2
              ICELNOD(4,ICELL2)=INODEND1+1
              ICELNOD(1,ICELL3)=INODEND1+1
              ICELNOD(2,ICELL3)=INODEND1+2
              ICELNOD(4,ICELL3)=INODEND1+3
              ICELNOD(2,ICELL4)=INODEND1+1
              ICELNOD(3,ICELL4)=INODEND1+3
C
              INODEND1=INODEND1+3
C
C.........  Get node from cell to south.
C
              IF(IFLAG(ICELLS).EQ.2) THEN
                  ICELNOD(2,ICELL1)=ICELNOD(3,ICELLS)
                  ICELNOD(1,ICELL2)=ICELNOD(2,ICELL1)
              ENDIF
C
C.........  Get node from cell to west.
C
              IF(IFLAG(ICELLW).EQ.2) THEN
                  ICELNOD(4,ICELL1)=ICELNOD(3,ICELLW)
                  ICELNOD(1,ICELL4)=ICELNOD(4,ICELL1)
              ENDIF
C
        ENDIF
C
C.........  Fix nodes on cells to north and east.
C
        IF(IFLAG(ICELLN).EQ.3) THEN
            ICELNOD(2,ICELLN)=ICELNOD(4,ICELL3)
            IFACE=ICELFAC(2,ICELLN)
            DO I=ICELBEG,ICELEND
                IF(ICELFAC(4,I).EQ.IFACE) THEN
                    ICELLT=I
                    GO TO 5
                ENDIF
            ENDDO
  5         CONTINUE
            ICELNOD(1,ICELLT)=ICELNOD(2,ICELLN)
        ENDIF
C
        IF(IFLAG(ICELLE).EQ.3) THEN
            ICELNOD(4,ICELLE)=ICELNOD(3,ICELL2)
```

```
          IFACE=ICELFAC(3,ICELLE)
          DO I=ICELBEG,ICELEND
             IF(ICELFAC(1,I).EQ.IFACE) THEN
                ICELLT=I
                GO TO 6
             ENDIF
          ENDDO
 6        CONTINUE
          ICELNOD(1,ICELLT)=ICELNOD(4,ICELLE)
      ENDIF
C
C.......... Cell-to-face pointer.
C
      IFACESB=ICELFAC(1,ICELL)
      IFACEEB=ICELFAC(2,ICELL)
      IFACENB=ICELFAC(3,ICELL)
      IFACEWB=ICELFAC(4,ICELL)
C
C.......... Southwest cell.
C
      IFACES=IFACESB
      IFACEN=IFA1END1+1
      IFACEW=IFACEWB
      IFACEE=IFA2END1+1
C
      ICELFAC(1,ICELL1)=IFACES
      ICELFAC(2,ICELL1)=IFACEE
      ICELFAC(3,ICELL1)=IFACEN
      ICELFAC(4,ICELL1)=IFACEW
C
C.......... Southeast cell.
C
      IFACES=IFA1END1+2
      IFACEN=IFA1END1+3
      IFACEW=IFA2END1+1
      IFACEE=IFACEEB
C
      ICELFAC(1,ICELL2)=IFACES
      ICELFAC(2,ICELL2)=IFACEE
      ICELFAC(3,ICELL2)=IFACEN
      ICELFAC(4,ICELL2)=IFACEW
C
C.......... Northeast cell.
C
      IFACES=IFA1END1+3
      IF(IFLAG(ICELLN).EQ.3) THEN
         IFACE=ICELFAC(2,ICELLN)
         DO I=ICELBEG,ICELEND
```

```
              IF(ICELFAC(4,I).EQ.IFACE) THEN
                  ICELLT=I
                  GO TO 7
              ENDIF
          ENDDO
7         CONTINUE
          IFACEN=ICELFAC(1,ICELLT)
      ELSE
          IFACEN=IFA1END1+4
      ENDIF
      IFACEW=IFA2END1+2
      IF(IFLAG(ICELLE).EQ.3) THEN
          IFACE=ICELFAC(3,ICELLE)
          DO I=ICELBEG,ICELEND
              IF(ICELFAC(1,I).EQ.IFACE) THEN
                  ICELLT=I
                  GO TO 8
              ENDIF
          ENDDO
8         CONTINUE
          IFACEE=ICELFAC(4,ICELLT)
      ELSE
          IFACEE=IFA2END1+4
      ENDIF
C
      ICELFAC(1,ICELL3)=IFACES
      ICELFAC(2,ICELL3)=IFACEE
      ICELFAC(3,ICELL3)=IFACEN
      ICELFAC(4,ICELL3)=IFACEW
C
C.......... Northwest cell.
C
      IFACES=IFA1END1+1
      IFACEN=IFACENB
      IFACEW=IFA2END1+3
      IFACEE=IFA2END1+2
C
      ICELFAC(1,ICELL4)=IFACES
      ICELFAC(2,ICELL4)=IFACEE
      ICELFAC(3,ICELL4)=IFACEN
      ICELFAC(4,ICELL4)=IFACEW
C
C.......... Fix faces on cells to south and west.
C
      IF(IFLAG(ICELLW).EQ.2) THEN
          IFACE=ICELFAC(3,ICELLW)
          DO I=ICELBEG,ICELEND
              IF(ICELFAC(1,I).EQ.IFACE) THEN
```

```
                  ICELLT=I
                  GO TO 9
              ENDIF
          ENDDO
 9        CONTINUE
          ICELFAC(2,ICELLT)=ICELFAC(4,ICELL4)
      ENDIF
C
      IF(IFLAG(ICELLS).EQ.2) THEN
          IFACE=ICELFAC(2,ICELLS)
          DO I=ICELBEG,ICELEND
              IF(ICELFAC(4,I).EQ.IFACE) THEN
                  ICELLT=I
                  GO TO 10
              ENDIF
          ENDDO
 10       CONTINUE
          ICELFAC(3,ICELLT)=ICELFAC(1,ICELL2)
      ENDIF
C
C.......... Update temporary pointers.
C
      IF(IFLAG(ICELLN).NE.0) THEN
          IFA1END1=IFA1END1+3
      ELSE
          IFA1END1=IFA1END1+4
      ENDIF
      IF(IFLAG(ICELLE).NE.0) THEN
          IFA2END1=IFA2END1+3
      ELSE
          IFA2END1=IFA2END1+4
      ENDIF
C
C.......... Update cell flags.
C
      IFLAG(ICELL1)=3
      IFLAG(ICELL2)=2
      IFLAG(ICELL3)=2
      IFLAG(ICELL4)=2
C
C
C
      RETURN
      END
```

```fortran
      SUBROUTINE DIVGRD(ILEM,ICELL,IFLAG,INODEND1,ICELEND1)
C
C$$$$$$$$$$$$$$$$$$$$$$$$$$$$$$$$$$$$$$$$$$$$$$$$$$$$$$$$$$$$$$$$$
C$$$$$$$$$$$$$$$$$$$$$$$$$$$$$$$$$$$$$$$$$$$$$$$$$$$$$$$$$$$$$$$$$
C$$$$$                                                       $$$$$
C$$$$$    This subroutine divides a cell, adding nodes to grid.  $$$$$
C$$$$$    Temporary node and cell pointers are used.         $$$$$
C$$$$$                                                       $$$$$
C$$$$$$$$$$$$$$$$$$$$$$$$$$$$$$$$$$$$$$$$$$$$$$$$$$$$$$$$$$$$$$$$$
C$$$$$$$$$$$$$$$$$$$$$$$$$$$$$$$$$$$$$$$$$$$$$$$$$$$$$$$$$$$$$$$$$
C
      INCLUDE 'LEVIS.INC'
      INTEGER IFLAG(NNODESMX)
      IDUMMY=NNODESMX
C
C.........  Find cell to south.
C
      ICELLS=IDUMMY
      IFACE=ICELFAC(1,ICELL)
      DO I=INODBEG,INODEND
         IF(ICELFAC(3,I).EQ.IFACE) THEN
             ICELLS=I
             GO TO 1
         ENDIF
      ENDDO
 1    CONTINUE
C
C.........  Find cell to north.
C
      ICELLN=IDUMMY
      IFACE=ICELFAC(3,ICELL)
      DO I=INODBEG,INODEND
         IF(ICELFAC(1,I).EQ.IFACE) THEN
             ICELLN=I
             GO TO 2
         ENDIF
      ENDDO
 2    CONTINUE
C
C.........  Find cell to west.
C
      ICELLW=IDUMMY
      IFACE=ICELFAC(4,ICELL)
      DO I=INODBEG,INODEND
         IF(ICELFAC(2,I).EQ.IFACE) THEN
             ICELLW=I
             GO TO 3
         ENDIF
      ENDDO
```

```
          ENDDO
 3       CONTINUE
C
C.........  Find cell to east.
C
         ICELLE=IDUMMY
         IFACE=ICELFAC(2,ICELL)
         DO I=INODBEG,INODEND
            IF(ICELFAC(4,I).EQ.IFACE) THEN
               ICELLE=I
               GO TO 4
            ENDIF
         ENDDO
 4       CONTINUE
C
C.........  Check for type of cell and add nodes.
C
         IF(IFLAG(ICELLS).EQ.0.AND.IFLAG(ICELLW).EQ.0) THEN
C
            ICELL1=ICELL
            ICELL2=ICELEND1+1
            ICELL3=ICELEND1+2
            ICELL4=ICELEND1+3
C
            ISW=ICELNOD(1,ICELL1)
            ISE=ICELNOD(2,ICELL2)
            INE=ICELNOD(3,ICELL3)
            INW=ICELNOD(4,ICELL4)
C
            X1(INODEND1+1)=0.5*(X1(ISW)+X1(ISE))
            X1(INODEND1+2)=0.25*(X1(ISW)+X1(ISE)+X1(INE)+X1(INW))
            X1(INODEND1+3)=0.5*(X1(ISW)+X1(INW))
            X1(INODEND1+4)=0.5*(X1(ISE)+X1(INE))
            X1(INODEND1+5)=0.5*(X1(INE)+X1(INW))
C
            X2(INODEND1+1)=0.5*(X2(ISW)+X2(ISE))
            X2(INODEND1+2)=0.25*(X2(ISW)+X2(ISE)+X2(INE)+X2(INW))
            X2(INODEND1+3)=0.5*(X2(ISW)+X2(INW))
            X2(INODEND1+4)=0.5*(X2(ISE)+X2(INE))
            X2(INODEND1+5)=0.5*(X2(INE)+X2(INW))
C
            ICELEND1=ICELEND1+3
            INODEND1=INODEND1+5
C
         ELSEIF(IFLAG(ICELLS).EQ.0) THEN
C
            ICELL1=ICELL
            ICELL2=ICELEND1+1
```

```
              ICELL3=ICELEND1+2
              ICELL4=ICELEND1+3
C
              ISW=ICELNOD(1,ICELL1)
              ISE=ICELNOD(2,ICELL2)
              INE=ICELNOD(3,ICELL3)
              INW=ICELNOD(4,ICELL4)
C
              X1(INODEND1+1)=0.5*(X1(ISW)+X1(ISE))
              X1(INODEND1+2)=0.25*(X1(ISW)+X1(ISE)+X1(INE)+X1(INW))
              X1(INODEND1+3)=0.5*(X1(ISE)+X1(INE))
              X1(INODEND1+4)=0.5*(X1(INE)+X1(INW))
C
              X2(INODEND1+1)=0.5*(X2(ISW)+X2(ISE))
              X2(INODEND1+2)=0.25*(X2(ISW)+X2(ISE)+X2(INE)+X2(INW))
              X2(INODEND1+3)=0.5*(X2(ISE)+X2(INE))
              X2(INODEND1+4)=0.5*(X2(INE)+X2(INW))
C
              ICELEND1=ICELEND1+3
              INODEND1=INODEND1+4
C
         ELSEIF(IFLAG(ICELLW).EQ.0) THEN
C
              ICELL1=ICELL
              ICELL2=ICELEND1+1
              ICELL3=ICELEND1+2
              ICELL4=ICELEND1+3
C
              ISW=ICELNOD(1,ICELL1)
              ISE=ICELNOD(2,ICELL2)
              INE=ICELNOD(3,ICELL3)
              INW=ICELNOD(4,ICELL4)
C
              X1(INODEND1+1)=0.25*(X1(ISW)+X1(ISE)+X1(INE)+X1(INW))
              X1(INODEND1+2)=0.5*(X1(ISW)+X1(INW))
              X1(INODEND1+3)=0.5*(X1(ISE)+X1(INE))
              X1(INODEND1+4)=0.5*(X1(INE)+X1(INW))
C
              X2(INODEND1+1)=0.25*(X2(ISW)+X2(ISE)+X2(INE)+X2(INW))
              X2(INODEND1+2)=0.5*(X2(ISW)+X2(INW))
              X2(INODEND1+3)=0.5*(X2(ISE)+X2(INE))
              X2(INODEND1+4)=0.5*(X2(INE)+X2(INW))
C
              ICELEND1=ICELEND1+3
              INODEND1=INODEND1+4
C
         ELSE
C
```

```
          ICELL1=ICELL
          ICELL2=ICELEND1+1
          ICELL3=ICELEND1+2
          ICELL4=ICELEND1+3
C
          ISW=ICELNOD(1,ICELL1)
          ISE=ICELNOD(2,ICELL2)
          INE=ICELNOD(3,ICELL3)
          INW=ICELNOD(4,ICELL4)
C
          X1(INODEND1+1)=0.25*(X1(ISW)+X1(ISE)+X1(INE)+X1(INW))
          X1(INODEND1+2)=0.5*(X1(ISE)+X1(INE))
          X1(INODEND1+3)=0.5*(X1(INE)+X1(INW))
C
          X2(INODEND1+1)=0.25*(X2(ISW)+X2(ISE)+X2(INE)+X2(INW))
          X2(INODEND1+2)=0.5*(X2(ISE)+X2(INE))
          X2(INODEND1+3)=0.5*(X2(INE)+X2(INW))
C
          ICELEND1=ICELEND1+3
          INODEND1=INODEND1+3
C
      ENDIF
C
C
C
      RETURN
      END
```

```
      SUBROUTINE DOUBBC
C
C$$$$$$$$$$$$$$$$$$$$$$$$$$$$$$$$$$$$$$$$$$$$$$$$$$$$$$$$$$$$$$$$$$$$
C$$$$$$$$$$$$$$$$$$$$$$$$$$$$$$$$$$$$$$$$$$$$$$$$$$$$$$$$$$$$$$$$$$$$
C$$$$$                                                           $$$$$
C$$$$$    This subroutine enforces the doubling condition at     $$$$$
C$$$$$    the solid boundary.                                    $$$$$
C$$$$$                                                           $$$$$
C$$$$$$$$$$$$$$$$$$$$$$$$$$$$$$$$$$$$$$$$$$$$$$$$$$$$$$$$$$$$$$$$$$$$
C$$$$$$$$$$$$$$$$$$$$$$$$$$$$$$$$$$$$$$$$$$$$$$$$$$$$$$$$$$$$$$$$$$$$
C
      INCLUDE 'LEVIS.INC'
C
C.........  Loop over the number of wall nodes.
C
      DO I=1,NNODWL
C
C.........  Find the wall node.
C
         INODE=INODWL(I)
C
C.........  Enforce the doubling condition at the node.
C
         DR (INODE)=2*DR (INODE)
         DRU(INODE)=2*DRU(INODE)
         DRV(INODE)=2*DRV(INODE)
         DRW(INODE)=2*DRW(INODE)
         DRE(INODE)=2*DRE(INODE)
C
      ENDDO
C
C
C
      RETURN
      END
```

```
      SUBROUTINE FARFBC
C
C$$$$$$$$$$$$$$$$$$$$$$$$$$$$$$$$$$$$$$$$$$$$$$$$$$$$$$$$$$$$$$$$$$
C$$$$$$$$$$$$$$$$$$$$$$$$$$$$$$$$$$$$$$$$$$$$$$$$$$$$$$$$$$$$$$$$$$
C$$$$$                                                        $$$$$
C$$$$$   This subroutine enforces the boundary conditions at  $$$$$
C$$$$$   the far-field boundary.                              $$$$$
C$$$$$                                                        $$$$$
C$$$$$$$$$$$$$$$$$$$$$$$$$$$$$$$$$$$$$$$$$$$$$$$$$$$$$$$$$$$$$$$$$$
C$$$$$$$$$$$$$$$$$$$$$$$$$$$$$$$$$$$$$$$$$$$$$$$$$$$$$$$$$$$$$$$$$$
C
      INCLUDE 'LEVIS.INC'
C
C.........  Loop over the number of far-field boundary nodes.
C
      DO I=1,NNODFF
C
C.........  Find the far-field node.
C
         INODE=INODFF(I)
C
C.........  Zero the changes at the node.
C
         DR (INODE)=0
         DRU(INODE)=0
         DRV(INODE)=0
         DRW(INODE)=0
         DRE(INODE)=0
C
         DAMP1(INODE)=0
         DAMP2(INODE)=0
         DAMP3(INODE)=0
         DAMP4(INODE)=0
         DAMP5(INODE)=0
C
      ENDDO
C
C
C
      RETURN
      END
```

```
      SUBROUTINE FLAGIT(IEMLEV,IFLAG)
C
C$$$$$$$$$$$$$$$$$$$$$$$$$$$$$$$$$$$$$$$$$$$$$$$$$$$$$$$$$$$$$$$$$$$$
C$$$$$$$$$$$$$$$$$$$$$$$$$$$$$$$$$$$$$$$$$$$$$$$$$$$$$$$$$$$$$$$$$$$$
C$$$$$                                                          $$$$$
C$$$$$   This subroutine flags all cells that are to be divided. $$$$$
C$$$$$                                                          $$$$$
C$$$$$$$$$$$$$$$$$$$$$$$$$$$$$$$$$$$$$$$$$$$$$$$$$$$$$$$$$$$$$$$$$$$$
C$$$$$$$$$$$$$$$$$$$$$$$$$$$$$$$$$$$$$$$$$$$$$$$$$$$$$$$$$$$$$$$$$$$$
C
      INCLUDE 'LEVIS.INC'
      INTEGER IFLAG(NNODESMX)
      INTEGER ICELLB(NLEMMAX)
C
C......... Statement functions for base rectangle pointers.
C
      ICELL(I1,I2,NWE)=(I1-1)*(NWE  )+I2
C
C......... Initialize flags to zero.
C
      DO I=INODBEG,INODEND
         IFLAG(I)=0
      ENDDO
      IFLAG(NNODESMX)=0
C
C......... Loop over base coordinates.
C
      DO I1=NEMSNBEG(IEMLEV),NEMSNEND(IEMLEV)
      DO I2=NEMWEBEG(IEMLEV),NEMWEEND(IEMLEV)
         ICELLB(1)=ICELL(I1,I2,NCELWE)
         IFLAG(ICELLB(1))=1
C
C......... Find remaining cells inside base cell.
C
         ICELL1=ICELLB(1)
         DO I=1,2**(IEMLEV-1)-1
            IFACE=ICELFAC(3,ICELL1)
            DO ITMP=ICELBEG,ICELEND
               IF(ICELFAC(1,ITMP).EQ.IFACE) THEN
                  ICELL1=ITMP
                  IFLAG(ITMP)=1
                  GO TO 1
               ENDIF
            ENDDO
 1          CONTINUE
         ENDDO
C
C......... Find bottom row of cells inside base cell.
```

```
C
          DO IB=1,2**(IEMLEV-1)-1
              IFACE=ICELFAC(2,ICELLB(IB))
              DO ITMP=ICELBEG,ICELEND
                  IF(ICELFAC(4,ITMP).EQ.IFACE) THEN
                      ICELLB(IB+1)=ITMP
                      IFLAG(ITMP)=1
                      GO TO 2
                  ENDIF
              ENDDO
 2        CONTINUE
C
C......... Find remaining cells inside base cell.
C
          ICELL1=ICELLB(IB+1)
          DO I=1,2**(IEMLEV-1)-1
              IFACE=ICELFAC(3,ICELL1)
              DO ITMP=ICELBEG,ICELEND
                  IF(ICELFAC(1,ITMP).EQ.IFACE) THEN
                      ICELL1=ITMP
                      IFLAG(ITMP)=1
                      GO TO 3
                  ENDIF
              ENDDO
 3        CONTINUE
          ENDDO
        ENDDO
      ENDDO
      ENDDO
C
C
C
      RETURN
      END
```

```
      SUBROUTINE FLUXBC
C
C$$$$$$$$$$$$$$$$$$$$$$$$$$$$$$$$$$$$$$$$$$$$$$$$$$$$$$$$$$$$$$$$$$
C$$$$$$$$$$$$$$$$$$$$$$$$$$$$$$$$$$$$$$$$$$$$$$$$$$$$$$$$$$$$$$$$$$
C$$$$$                                                         $$$$$
C$$$$$   This subroutine enforces the no-flux condition. It is $$$$$
C$$$$$   done by zeroing the mass-flux through the faces on the $$$$$
C$$$$$   wall, keeping only the pressure terms.                $$$$$
C$$$$$                                                         $$$$$
C$$$$$$$$$$$$$$$$$$$$$$$$$$$$$$$$$$$$$$$$$$$$$$$$$$$$$$$$$$$$$$$$$$
C$$$$$$$$$$$$$$$$$$$$$$$$$$$$$$$$$$$$$$$$$$$$$$$$$$$$$$$$$$$$$$$$$$
C
      INCLUDE 'LEVIS.INC'
C
C.........  Statement functions for flux vectors.
C
      F1(I)=0
      F2(I)=         -X1(I)*P(I)
      F3(I)=            P(I)
      F4(I)=0
      F5(I)=0
C
      G1(I)=0
      G2(I)=         -X2(I)*P(I)
      G3(I)=0
      G4(I)=            P(I)
      G5(I)=0
C
C.........  Loop over the number of wall faces.
C
      DO I=1,NFACWL
C
C.........  Find the face on the wall.
C
         IFACE=IFACWL(I)
C
C.........  Find the nodes that define the face.
C
         INODE1=IFA1NOD(1,IFACE)
         INODE2=IFA1NOD(2,IFACE)
C
C.........  Calculate the normal.
C
         DX1=X1(INODE2)-X1(INODE1)
         DX2=X2(INODE2)-X2(INODE1)
C
C.........  Trapezoidal integration for the flux.
C
```

```
              FLUX1(1,IFACE)=.5*( (F1(INODE1)+F1(INODE2))*DX2 -
     &                            (G1(INODE1)+G1(INODE2))*DX1  )
              FLUX2(1,IFACE)=.5*( (F2(INODE1)+F2(INODE2))*DX2 -
     &                            (G2(INODE1)+G2(INODE2))*DX1  )
              FLUX3(1,IFACE)=.5*( (F3(INODE1)+F3(INODE2))*DX2 -
     &                            (G3(INODE1)+G3(INODE2))*DX1  )
              FLUX4(1,IFACE)=.5*( (F4(INODE1)+F4(INODE2))*DX2 -
     &                            (G4(INODE1)+G4(INODE2))*DX1  )
              FLUX5(1,IFACE)=.5*( (F5(INODE1)+F5(INODE2))*DX2 -
     &                            (G5(INODE1)+G5(INODE2))*DX1  )
C
          ENDDO
C
C
C
          RETURN
          END
```

```
      SUBROUTINE GENBND
C
C$$$$$$$$$$$$$$$$$$$$$$$$$$$$$$$$$$$$$$$$$$$$$$$$$$$$$$$$$$$$$$$$$$$$
C$$$$$$$$$$$$$$$$$$$$$$$$$$$$$$$$$$$$$$$$$$$$$$$$$$$$$$$$$$$$$$$$$$$$
C$$$$$                                                          $$$$$
C$$$$$    This subroutine sets up the boundary and interface    $$$$$
C$$$$$    pointers.                                             $$$$$
C$$$$$                                                          $$$$$
C$$$$$$$$$$$$$$$$$$$$$$$$$$$$$$$$$$$$$$$$$$$$$$$$$$$$$$$$$$$$$$$$$$$$
C$$$$$$$$$$$$$$$$$$$$$$$$$$$$$$$$$$$$$$$$$$$$$$$$$$$$$$$$$$$$$$$$$$$$
C
      INCLUDE 'LEVIS.INC'
C
      INODEND1=(NCELWE+1)*(NCELSN+1)
C
C.........  Set up west periodic boundary pointers.
C
      NNODPW=0
      INODPW(1)=INODBEG
      DO I=1,NNODESMX
         NNODPW=NNODPW+1
         INODPW(I+1)=INODNOD(3,INODPW(I))
         IF(INODPW(I+1).EQ.NNODESMX) GO TO 1
      ENDDO
 1    CONTINUE
C
C.........  Set up east periodic boundary pointers.
C
      NNODPE=0
      INODPE(1)=INODEND1
      DO I=1,NNODESMX
         NNODPE=NNODPE+1
         INODPE(I+1)=INODNOD(1,INODPE(I))
         IF(INODPE(I+1).EQ.NNODESMX) GO TO 2
      ENDDO
 2    CONTINUE
C
C.........  Set up far-field boundary pointers.
C
      NNODFF=0
      INODFF(1)=INODPW(NNODPW)
      DO I=1,NNODESMX
         NNODFF=NNODFF+1
         INODFF(I+1)=INODNOD(2,INODFF(I))
         IF(INODFF(I+1).EQ.NNODESMX) GO TO 3
      ENDDO
 3    CONTINUE
C
```

```
C.........  Set up wall pointers.
C
      NNODWL=0
      INODWL(1)=INODPE(NNODPE)
      DO I=1,NNODESMX
          NNODWL=NNODWL+1
          INODWL(I+1)=INODNOD(4,INODWL(I))
          IF(INODWL(I+1).EQ.NNODESMX) GO TO 4
      ENDDO
  4   CONTINUE
C
      NFACWL=0
      ICELL=ICELBEG
      DO I=1,NNODWL-1
          NFACWL=NFACWL+1
          IFACWL(I)=ICELFAC(1,ICELL)
          CALL CELCEL(2,ICELL,ICELLT)
          ICELL=ICELLT
      ENDDO
C
C.........  Embedded interface pointers.
C
      NCELINT=0
      DO ILEM=1,NLEM
C
C.........  Cells to south of an interface.
C
          IF(NEMSNBEG(ILEM).NE.1) THEN
              ICELLSW=(NEMSNBEG(ILEM)-1)*NCELWE+NEMWEBEG(ILEM)
              CALL CELCEL(2,ICELLSW,ICELLSE)
              CALL CELCEL(1,ICELLSW,ICELL)
              DO J=1,NEMWE(ILEM)*2**(ILEM-1)
                  NCELINT=NCELINT+1
                  ICELINT(NCELINT)=ICELL
                  ICELFACI(1,NCELINT)=ICELFAC(1,ICELLSE)
                  ICELFACI(2,NCELINT)=3
                  ICELNODI(NCELINT)=ICELNOD(2,ICELLSW)
                  IFA1NOD(2,ICELFAC(1,ICELLSW))=ICELNODI(NCELINT)
                  INODNOD(2,ICELNOD(1,ICELLSW))=ICELNODI(NCELINT)
                  INODNOD(4,ICELNOD(2,ICELLSE))=ICELNODI(NCELINT)
                  CALL CELCEL(2,ICELL,ICELLT)
                  ICELL=ICELLT
                  CALL CELCEL(3,ICELL,ICELLSW)
                  CALL CELCEL(2,ICELLSW,ICELLSE)
              ENDDO
          ENDIF
C
C.........  Cells to west of an interface.
```

```
        C
                IF(NEMWEBEG(ILEM).NE.1) THEN
                    ICELLSW=(NEMSNBEG(ILEM)-1)*NCELWE+NEMWEBEG(ILEM)
                    CALL CELCEL(3,ICELLSW,ICELLNW)
                    CALL CELCEL(4,ICELLSW,ICELL)
                    DO J=1,NEMSN(ILEM)*2**(ILEM-1)
                        NCELINT=NCELINT+1
                        ICELINT(NCELINT)=ICELL
                        ICELFACI(1,NCELINT)=ICELFAC(4,ICELLNW)
                        ICELFACI(2,NCELINT)=2
                        ICELNODI(NCELINT)=ICELNOD(4,ICELLSW)
                        IFA2NOD(2,ICELFAC(4,ICELLSW))=ICELNODI(NCELINT)
                        INODNOD(3,ICELNOD(1,ICELLSW))=ICELNODI(NCELINT)
                        INODNOD(1,ICELNOD(4,ICELLNW))=ICELNODI(NCELINT)
                        CALL CELCEL(3,ICELL,ICELLT)
                        ICELL=ICELLT
                        CALL CELCEL(2,ICELL,ICELLSW)
                        CALL CELCEL(3,ICELLSW,ICELLNW)
                    ENDDO
                ENDIF
        C
        C.......... Cells to north of an interface.
        C
                IF(NEMSNEND(ILEM).NE.NCELSN) THEN
                    ICELLSW=(NEMSNEND(ILEM)-1)*NCELWE+NEMWEEND(ILEM)
                    DO J=1,2**ILEM-2
                        CALL CELCEL(2,ICELLSW,ICELLT)
                        CALL CELCEL(3,ICELLT,ICELLSW)
                    ENDDO
                    CALL CELCEL(3,ICELLSW,ICELLNW)
                    CALL CELCEL(2,ICELLNW,ICELLNE)
                    CALL CELCEL(3,ICELLNW,ICELL)
                    DO J=1,NEMWE(ILEM)*2**(ILEM-1)
                        NCELINT=NCELINT+1
                        ICELINT(NCELINT)=ICELL
                        ICELFACI(1,NCELINT)=ICELFAC(3,ICELLNE)
                        ICELFACI(2,NCELINT)=1
                        ICELNODI(NCELINT)=ICELNOD(3,ICELLNW)
                        IFA1NOD(2,ICELFAC(3,ICELLNW))=ICELNODI(NCELINT)
                        INODNOD(2,ICELNOD(4,ICELLNW))=ICELNODI(NCELINT)
                        INODNOD(4,ICELNOD(3,ICELLNE))=ICELNODI(NCELINT)
                        CALL CELCEL(4,ICELL,ICELLT)
                        ICELL=ICELLT
                        CALL CELCEL(1,ICELL,ICELLNW)
                        CALL CELCEL(2,ICELLNW,ICELLNE)
                    ENDDO
                ENDIF
        C
```

```
C.......... Cells to east of an interface.
C
          IF(NEMWEEND(ILEM).NE.NCELWE) THEN
              ICELLSW=(NEMSNEND(ILEM)-1)*NCELWE+NEMWEEND(ILEM)
              DO J=1,2**ILEM-2
                  CALL CELCEL(2,ICELLSW,ICELLT)
                  CALL CELCEL(3,ICELLT,ICELLSW)
              ENDDO
              CALL CELCEL(2,ICELLSW,ICELLSE)
              CALL CELCEL(3,ICELLSE,ICELLNE)
              CALL CELCEL(2,ICELLSE,ICELL)
              DO J=1,NEMSN(ILEM)*2**(ILEM-1)
                  NCELINT=NCELINT+1
                  ICELINT(NCELINT)=ICELL
                  ICELFACI(1,NCELINT)=ICELFAC(2,ICELLNE)
                  ICELFACI(2,NCELINT)=4
                  ICELNODI(NCELINT)=ICELNOD(3,ICELLSE)
                  IFA2NOD(2,ICELFAC(2,ICELLSE))=ICELNODI(NCELINT)
                  INODNOD(3,ICELNOD(2,ICELLSE))=ICELNODI(NCELINT)
                  INODNOD(1,ICELNOD(3,ICELLNE))=ICELNODI(NCELINT)
                  CALL CELCEL(1,ICELL,ICELLT)
                  ICELL=ICELLT
                  CALL CELCEL(4,ICELL,ICELLSE)
                  CALL CELCEL(3,ICELLSE,ICELLNE)
              ENDDO
          ENDIF
      ENDDO
C
C.......... Set up smoothing override pointers near leading-edge.
C
      NNODOV=8
      INODOV(1)=NCELWE/2+1
      INODOV(2)=INODNOD(2,INODOV(1))
      INODOV(3)=INODNOD(3,INODOV(1))
      INODOV(4)=INODNOD(4,INODOV(1))
      INODOV(5)=INODNOD(3,INODOV(2))
      INODOV(6)=INODNOD(3,INODOV(4))
      INODOV(7)=INODNOD(4,INODOV(4))
      INODOV(8)=INODNOD(3,INODOV(7))
C
C.......... Set up node-to-node pointer on boundaries.
C
      DO I=1,NNODPW
          INODNOD(4,INODPW(I))=NNODESMX
      ENDDO
      DO I=1,NNODPE
          INODNOD(2,INODPE(I))=NNODESMX
      ENDDO
```

```
      DO I=1,NNODFF
          INODNOD(3,INODFF(I))=NNODESMX
      ENDDO
      DO I=1,NNODWL
          INODNOD(1,INODWL(I))=NNODESMX
      ENDDO
C
C.......... Set up damping pointer for number of cells about node.
C
      DO I=INODBEG,INODEND
          NCN(I)=4
      ENDDO
      DO IT=1,NNODWL
          I=INODWL(IT)
          NCN(I)=NCN(I)/2
      ENDDO
      DO IT=1,NNODFF
          I=INODFF(IT)
          NCN(I)=NCN(I)/2
      ENDDO
      DO IT=1,NNODPW
          I=INODPW(IT)
          NCN(I)=NCN(I)/2
      ENDDO
      DO IT=1,NNODPE
          I=INODPE(IT)
          NCN(I)=NCN(I)/2
      ENDDO
      DO IT=1,NCELINT
          I=ICELNODI(IT)
          NCN(I)=NCN(I)/2
      ENDDO
C
C
C
      RETURN
      END
```

```
      SUBROUTINE GENEMB
C
C$$$$$$$$$$$$$$$$$$$$$$$$$$$$$$$$$$$$$$$$$$$$$$$$$$$$$$$$$$$$$$$$$$$$
C$$$$$$$$$$$$$$$$$$$$$$$$$$$$$$$$$$$$$$$$$$$$$$$$$$$$$$$$$$$$$$$$$$$$
C$$$$$                                                          $$$$$
C$$$$$    This subroutine fixes the pointers and grid for the   $$$$$
C$$$$$    embedded region.                                      $$$$$
C$$$$$                                                          $$$$$
C$$$$$$$$$$$$$$$$$$$$$$$$$$$$$$$$$$$$$$$$$$$$$$$$$$$$$$$$$$$$$$$$$$$$
C$$$$$$$$$$$$$$$$$$$$$$$$$$$$$$$$$$$$$$$$$$$$$$$$$$$$$$$$$$$$$$$$$$$$
C
      INCLUDE 'LEVIS.INC'
      INTEGER IFLAG(NNODESMX)
C
C.......... Length of embedded regions.
C
      DO I=1,NLEM
         IF(NEMWEBEG(I).NE.0.AND.NEMWEEND(I).NE.0.AND.
     &      NEMSNBEG(I).NE.0.AND.NEMSNEND(I).NE.0)  THEN
            NEMWE(I)=NEMWEEND(I)-NEMWEBEG(I)+1
            NEMSN(I)=NEMSNEND(I)-NEMSNBEG(I)+1
         ELSE
            NEMWE(I)=0
            NEMSN(I)=0
            NEMSNBEG(I)=0
            NEMWEBEG(I)=0
            NEMSNEND(I)=-1
            NEMWEEND(I)=-1
         ENDIF
      ENDDO
C
C.......... Pointers for primary recatngle.
C
      ICELEND1=(NCELWE  )*(NCELSN  )
      INODEND1=(NCELWE+1)*(NCELSN+1)
      IFA1END1=(NCELWE  )*(NCELSN+1)
      IFA2END1=(NCELWE+1)*(NCELSN  )
C
C.......... Add for embedded region.
C
      DO I=1,NLEM
         ICELEND=ICELEND+3*4**(I-1)*NEMWE(I)*NEMSN(I)
         INODEND=INODEND+3*4**(I-1)*NEMWE(I)*NEMSN(I)+
     &                   2**(I-1)*NEMWE(I)+
     &                   2**(I-1)*NEMSN(I)
         IFA1END=IFA1END+3*4**(I-1)*NEMWE(I)*NEMSN(I)+
     &                   2**(I-1)*NEMWE(I)
         IFA2END=IFA2END+3*4**(I-1)*NEMWE(I)*NEMSN(I)+
```

```
      &                      2**(I-1)*NEMSN(I)
      ENDDO
C
      IF(INODEND.GE.NNODESMX) THEN
          WRITE(6,*) 'Too many nodes'
          STOP
      ENDIF
C
C.........  Fix pointers in embedded regions.
C
      DO I=1,NLEM
C
C.........  Flag all cells to be divided at this level.
C
          CALL FLAGIT(I,IFLAG)
C
C.........  Divide all flagged cells.
C
          DO ICELL=ICELBEG,ICELEND
              IF(IFLAG(ICELL).EQ.1) THEN
                  ICELS=ICELEND1
                  INODS=INODEND1
                  CALL DIVCEL(I,ICELL,IFLAG,
     &                    INODEND1,ICELEND1,IFA1END1,IFA2END1)
                  ICELEND1=ICELS
                  INODEND1=INODS
              ENDIF
              IF(IFLAG(ICELL).EQ.3) THEN
                  CALL DIVGRD(I,ICELL,IFLAG,INODEND1,ICELEND1)
              ENDIF
          ENDDO
      ENDDO
C
C.........  Node-to-node pointer.
C
      DO I=ICELBEG,ICELEND
C
          ISW=ICELNOD(1,I)
          ISE=ICELNOD(2,I)
          INE=ICELNOD(3,I)
          INW=ICELNOD(4,I)
C
          INODNOD(1,INE)=ISE
          INODNOD(1,INW)=ISW
          INODNOD(2,INW)=INE
          INODNOD(2,ISW)=ISE
          INODNOD(3,ISE)=INE
          INODNOD(3,ISW)=INW
```

```
            INODNOD(4,INE)=INW
            INODNOD(4,ISE)=ISW
C
      ENDDO
C
C.........  Face-to-node pointer.
C
      DO I=ICELBEG,ICELEND
C
            IFACES=ICELFAC(1,I)
            IFACEE=ICELFAC(2,I)
            IFACEN=ICELFAC(3,I)
            IFACEW=ICELFAC(4,I)
C
            ISW=ICELNOD(1,I)
            ISE=ICELNOD(2,I)
            INE=ICELNOD(3,I)
            INW=ICELNOD(4,I)
C
            IFA1NOD(1,IFACES)=ISW
            IFA1NOD(2,IFACES)=ISE
            IFA1NOD(1,IFACEN)=INW
            IFA1NOD(2,IFACEN)=INE
C
            IFA2NOD(1,IFACEW)=ISW
            IFA2NOD(2,IFACEW)=INW
            IFA2NOD(1,IFACEE)=ISE
            IFA2NOD(2,IFACEE)=INE
C
      ENDDO
C
C
C
      RETURN
      END
```

```
      SUBROUTINE GENGRD
C
C$$$$$$$$$$$$$$$$$$$$$$$$$$$$$$$$$$$$$$$$$$$$$$$$$$$$$$$$$$$$$$$$$$
C$$$$$$$$$$$$$$$$$$$$$$$$$$$$$$$$$$$$$$$$$$$$$$$$$$$$$$$$$$$$$$$$$$
C$$$$$                                                         $$$$$
C$$$$$   This routine generates a grid for a flat plate geometry. $$$$$
C$$$$$                                                         $$$$$
C$$$$$$$$$$$$$$$$$$$$$$$$$$$$$$$$$$$$$$$$$$$$$$$$$$$$$$$$$$$$$$$$$$
C$$$$$$$$$$$$$$$$$$$$$$$$$$$$$$$$$$$$$$$$$$$$$$$$$$$$$$$$$$$$$$$$$$
C
      INCLUDE 'LEVIS.INC'
      COMPLEX ZETA,ZETABOD,ZETATMP,ZETAFAR
C
C.........  Major and minor axes of wing
C
      PI=2.0*ASIN(1.0)
      ECC=1.0
      A  = TAN(CONEANG)
      B  =A*SQRT(1-ECC**2)
C
C.........  Parameters for Joukowski transformation.
C
      C=.5*(A+B)
      S=.5*SQRT(A**2-B**2)
C
C.........  Loop over rays.
C
      DO J=1,NCELWE+1
C
C.........  Generate transformed body points.
C
          T=PI*(J-1)/NCELWE
C
          ZETABOD=C*CMPLX(SIN(T),COS(T))
C
C.........  Transform outer boundary points.
C
          RSHK=3.0*ASIN(1.0/MINF)
          OFFSET=MIN(3.0*SIN(ALPHA),3.0*SIN(10.*PI/180.))
          XSHK=RSHK*SIN(T)
          YSHK=RSHK*COS(T)+OFFSET*COS(T)**2
          ZETATMP=CMPLX(XSHK,YSHK)
C
C.........  Fix boundaries to avoid alliant complex square root bug.
C
          IF(J.EQ.1) THEN
              ZETABOD=CMPLX(0.0,C)
              YSHK=RSHK+OFFSET
```

```
              ZETATMP=CMPLX(0.0,YSHK)
           ELSEIF(J.EQ.NCELWE+1) THEN
              ZETABOD=CMPLX(0.0,-C)
              YSHK=-RSHK+OFFSET
              ZETATMP=CMPLX(0.0,YSHK)
           ENDIF
C
C.........  Choose root by quadrant of shock point.
C
           IF(J.NE.NCELWE+1) THEN
              ZETAFAR = .5*ZETATMP+.5*SQRT(ZETATMP**2-4*S**2)
           ELSE
              ZETAFAR = .5*ZETATMP-.5*SQRT(ZETATMP**2-4*S**2)
           ENDIF
C
C.........  Divide ray into segments, with exponential stretching.
C
           DO I=1,NCELSN+1
              COMPX=(I-1.)/NCELSN
              RBAR=-LOG(1.-(1.-EXP(-SIGMA))*COMPX)/SIGMA
              ZETATMP=ZETABOD+(ZETAFAR-ZETABOD)*RBAR
              ZETA=ZETATMP+(S**2)/ZETATMP
C
C.........  Store grid.
C
              INODE=(I-1)*(NCELWE+1)+J
              X1(INODE)= REAL(ZETA)
              X2(INODE)=AIMAG(ZETA)
C
           ENDDO
        ENDDO
C
C
C
      RETURN
      END
```

```
      SUBROUTINE GENPNT
C
C$$$$$$$$$$$$$$$$$$$$$$$$$$$$$$$$$$$$$$$$$$$$$$$$$$$$$$$$$$$$$$$$$$$$$
C$$$$$$$$$$$$$$$$$$$$$$$$$$$$$$$$$$$$$$$$$$$$$$$$$$$$$$$$$$$$$$$$$$$$$
C$$$$$                                                           $$$$$
C$$$$$     This subroutine generates the pointers for a logically $$$$$
C$$$$$     rectangular symmetric flat plate grid.                $$$$$
C$$$$$                                                           $$$$$
C$$$$$$$$$$$$$$$$$$$$$$$$$$$$$$$$$$$$$$$$$$$$$$$$$$$$$$$$$$$$$$$$$$$$$
C$$$$$$$$$$$$$$$$$$$$$$$$$$$$$$$$$$$$$$$$$$$$$$$$$$$$$$$$$$$$$$$$$$$$$
C
      INCLUDE 'LEVIS.INC'
C
C.........  Statement functions for pointers.
C
      ISN(I,NWE)=(I-1)/(NWE+1)+1
      IWE(I,NWE)=MOD(I-1,NWE+1)+1
      INODE(I1,I2,NWE)=(I1-1)*(NWE+1)+I2
C
C.........  Set up IDUMMY pointer.
C
      IDUMMY=NNODESMX
C
C.........  Number of cells defining primary rectangle.
C
      NWE=NCELWE
      NSN=NCELSN
C
C.........  Set up bounds for number of cells, faces and nodes.
C
      INODBEG=1
      IFA1BEG=1
      IFA2BEG=1
      ICELBEG=1
C
      ICELEND1=ICELBEG+(NWE  )*(NSN  )-1
      INODEND1=INODBEG+(NWE+1)*(NSN+1)-1
      IFA1END1=IFA1BEG+(NWE  )*(NSN+1)-1
      IFA2END1=IFA2BEG+(NWE+1)*(NSN  )-1
C
      ICELEND=ICELEND1
      INODEND=INODEND1
      IFA1END=IFA1END1
      IFA2END=IFA2END1
C
C.........  Set up pointers for primary rectangle.
C
      DO I=INODBEG,INODEND
```

```
C
C.......... Set up south, east, north and west node-to-node pointers.
C
          INODNOD(1,I)=INODE(ISN(I,NWE)-1,
     &                      IWE(I,NWE)   ,NWE)
          INODNOD(2,I)=INODE(ISN(I,NWE)  ,
     &                      IWE(I,NWE)+1,NWE)
          INODNOD(3,I)=INODE(ISN(I,NWE)+1,
     &                      IWE(I,NWE)   ,NWE)
          INODNOD(4,I)=INODE(ISN(I,NWE)  ,
     &                      IWE(I,NWE)-1,NWE)
C
C.......... Set any artificially low nodes to IDUMMY.
C
          IF(INODNOD(1,I).LT.1) INODNOD(1,I)=IDUMMY
          IF(INODNOD(2,I).LT.1) INODNOD(2,I)=IDUMMY
          IF(INODNOD(3,I).LT.1) INODNOD(3,I)=IDUMMY
          IF(INODNOD(4,I).LT.1) INODNOD(4,I)=IDUMMY
C
C.......... Set any artificially high nodes to IDUMMY.
C
          IF(INODNOD(1,I).GT.(NWE+1)*(NSN+1)) INODNOD(1,I)=IDUMMY
          IF(INODNOD(2,I).GT.(NWE+1)*(NSN+1)) INODNOD(2,I)=IDUMMY
          IF(INODNOD(3,I).GT.(NWE+1)*(NSN+1)) INODNOD(3,I)=IDUMMY
          IF(INODNOD(4,I).GT.(NWE+1)*(NSN+1)) INODNOD(4,I)=IDUMMY
C
C.......... Set dummy pointers to dummy node.
C
          INODNOD(1,IDUMMY)=IDUMMY
          INODNOD(2,IDUMMY)=IDUMMY
          INODNOD(3,IDUMMY)=IDUMMY
          INODNOD(4,IDUMMY)=IDUMMY
C
C.......... Define the north-south contravariant face.
C
          IFACES=I-ISN(I,NWE)+1
          IF(IFACE.GT.IFA1END) IFACE=IDUMMY
          IF(IFACE.LT.IFA1BEG) IFACE=IDUMMY
          IFA1NOD(1,IFACES)=I
          IFA1NOD(2,IFACES)=I+1
C
C.......... Define the west-east contravariant face.
C
          IFACEW=I
          IFA2NOD(1,IFACEW)=I
          IFA2NOD(2,IFACEW)=I+NWE+1
C
C.......... Define the cell to the northeast.
```

```
C
         ICELL=I-ISN(I,NWE)+1
         ICELNOD(1,ICELL)=I
         ICELNOD(2,ICELL)=I+1
         ICELNOD(3,ICELL)=I+NWE+2
         ICELNOD(4,ICELL)=I+NWE+1
C
C.........  Construct the cell-to-face pointer.
C
         ICELFAC(1,ICELL)=IFACES
         ICELFAC(2,ICELL)=IFACEW+1
         ICELFAC(3,ICELL)=IFACES+NWE
         ICELFAC(4,ICELL)=IFACEW
C
      ENDDO
C
C
C
      RETURN
      END
```

```
      SUBROUTINE OUTPUT
C
C$$$$$$$$$$$$$$$$$$$$$$$$$$$$$$$$$$$$$$$$$$$$$$$$$$$$$$$$$$$$$$$$$$$$
C$$$$$$$$$$$$$$$$$$$$$$$$$$$$$$$$$$$$$$$$$$$$$$$$$$$$$$$$$$$$$$$$$$$$
C$$$$$                                                          $$$$$
C$$$$$    This subroutine creates the output file.              $$$$$
C$$$$$                                                          $$$$$
C$$$$$$$$$$$$$$$$$$$$$$$$$$$$$$$$$$$$$$$$$$$$$$$$$$$$$$$$$$$$$$$$$$$$
C$$$$$$$$$$$$$$$$$$$$$$$$$$$$$$$$$$$$$$$$$$$$$$$$$$$$$$$$$$$$$$$$$$$$
C
C
      INCLUDE 'LEVIS.INC'
C
C.........  Write the file.
C
      PI=2.0*ASIN(1.0)
      OPEN(UNIT=1,STATUS='NEW',FORM='UNFORMATTED',FILE='LEVISOUT.DAT')
      ISYM=1
      NWEEQUIV=NCELWE*2**NLEM
      NSNEQUIV=NCELSN*2**NLEM
      WRITE(1) ICELEND,INODEND,ISYM
      WRITE(1) MINF,EPS2,EPS4,NWEEQUIV,NSNEQUIV
      WRITE(1) ALPHA*180/PI,YAW*180/PI,90-CONEANG*180/PI
      DO I=INODBEG,INODEND
          WRITE(1) X1(I),X2(I)
      ENDDO
      DO I=INODBEG,INODEND
          WRITE(1) RHO(I),RHOU(I),RHOV(I),RHOW(I),RHOE(I),DP(I)
      ENDDO
      DO I=ICELBEG,ICELEND
          WRITE(1) ICELNOD(1,I)
      ENDDO
      DO I=ICELBEG,ICELEND
          WRITE(1) ICELNOD(2,I)
      ENDDO
      DO I=ICELBEG,ICELEND
          WRITE(1) ICELNOD(3,I)
      ENDDO
      DO I=ICELBEG,ICELEND
          WRITE(1) ICELNOD(4,I)
      ENDDO
      WRITE(1) NNODWL
      WRITE(1) NNODFF
      WRITE(1) NNODPW
      WRITE(1) NNODPE
      DO I=1,NNODWL
          WRITE(1) INODWL(I)
      ENDDO
```

```
      DO I=1,NNODFF
          WRITE(1) INODFF(I)
      ENDDO
      DO I=1,NNODPW
          WRITE(1) INODPW(I)
      ENDDO
      DO I=1,NNODPE
          WRITE(1) INODPE(I)
      ENDDO
      WRITE(1) NCELINT
      DO I=1,NCELINT
          WRITE(1) ICELINT(I),ICELFACI(2,I)
      ENDDO
C
C.......... Close output file.
C
      CLOSE (UNIT=1)
C
C
C
      RETURN
      END
```

```
      SUBROUTINE PERIBC
C
C$$$$$$$$$$$$$$$$$$$$$$$$$$$$$$$$$$$$$$$$$$$$$$$$$$$$$$$$$$$$$$$$$$$$$
C$$$$$$$$$$$$$$$$$$$$$$$$$$$$$$$$$$$$$$$$$$$$$$$$$$$$$$$$$$$$$$$$$$$$$
C$$$$$                                                           $$$$$
C$$$$$    This subroutine enforces the boundary conditions at    $$$$$
C$$$$$    the periodic boundary.                                 $$$$$
C$$$$$                                                           $$$$$
C$$$$$$$$$$$$$$$$$$$$$$$$$$$$$$$$$$$$$$$$$$$$$$$$$$$$$$$$$$$$$$$$$$$$$
C$$$$$$$$$$$$$$$$$$$$$$$$$$$$$$$$$$$$$$$$$$$$$$$$$$$$$$$$$$$$$$$$$$$$$
C
      INCLUDE 'LEVIS.INC'
C
C.........  Loop over the west periodic boundary nodes.
C
CVD$  NODEPCHK
      DO I=1,NNODPW
C
C.........  Find the periodic boundary node.
C
          INODE=INODPW(I)
C
C.........  Apply the boundary conditions.
C
          DR (INODE)=2*DR (INODE)
          DRU(INODE)=2*DRU(INODE)
          DRV(INODE)=0
          DRW(INODE)=2*DRW(INODE)
          DRE(INODE)=2*DRE(INODE)
C
          DAMP1(INODE)=2*DAMP1(INODE)
          DAMP2(INODE)=2*DAMP2(INODE)
          DAMP3(INODE)=0
          DAMP4(INODE)=2*DAMP4(INODE)
          DAMP5(INODE)=2*DAMP5(INODE)
C
      ENDDO
C
C.........  Loop over the east periodic boundary nodes.
C
CVD$  NODEPCHK
      DO I=1,NNODPE
C
C.........  Find the periodic boundary node.
C
          INODE=INODPE(I)
C
C.........  Apply the boundary conditions.
```

```
C
            DR (INODE)=2*DR(INODE)
            DRU(INODE)=2*DRU(INODE)
            DRV(INODE)=0
            DRW(INODE)=2*DRW(INODE)
            DRE(INODE)=2*DRE(INODE)
C
            DAMP1(INODE)=2*DAMP1(INODE)
            DAMP2(INODE)=2*DAMP2(INODE)
            DAMP3(INODE)=0
            DAMP4(INODE)=2*DAMP4(INODE)
            DAMP5(INODE)=2*DAMP5(INODE)
C
      ENDDO
C
C
C
      RETURN
      END
```

```
      SUBROUTINE READIN
C
C$$$$$$$$$$$$$$$$$$$$$$$$$$$$$$$$$$$$$$$$$$$$$$$$$$$$$$$$$$$$$$$$$$$
C$$$$$$$$$$$$$$$$$$$$$$$$$$$$$$$$$$$$$$$$$$$$$$$$$$$$$$$$$$$$$$$$$$$
C$$$$$                                                         $$$$$
C$$$$$    This subroutine reads the input parameters for LEVIS. $$$$$
C$$$$$                                                         $$$$$
C$$$$$$$$$$$$$$$$$$$$$$$$$$$$$$$$$$$$$$$$$$$$$$$$$$$$$$$$$$$$$$$$$$$
C$$$$$$$$$$$$$$$$$$$$$$$$$$$$$$$$$$$$$$$$$$$$$$$$$$$$$$$$$$$$$$$$$$$
C
      INCLUDE 'LEVIS.INC'
C
C.........  Read the input quantities.
C
      OPEN(UNIT=1,FILE='LEVISIN.DAT',STATUS='OLD')
      READ(1,*) NCELWE
      READ(1,*) NCELSN
      READ(1,*) SIGMA
      READ(1,*) MINF
      READ(1,*) ALPHAD
      READ(1,*) BETAD
      READ(1,*) SWEEPD
      READ(1,*) ITERS
      READ(1,*) CFL
      READ(1,*) EPS2
      READ(1,*) EPS4
      READ(1,*) RESTART
      READ(1,*) NLEM
      DO I=1,NLEM
          READ(1,*) NEMWEBEG(I)
          READ(1,*) NEMWEEND(I)
          READ(1,*) NEMSNBEG(I)
          READ(1,*) NEMSNEND(I)
      ENDDO
      CLOSE(UNIT=1)
      SYM=.TRUE.
C
C.........  Convert the angles to radians
C
      PI=2.0*ASIN(1.0)
      ALPHA=ALPHAD*PI/180.
      BETA=BETAD*PI/180.
      SWEEP=SWEEPD*PI/180.
      CONEANG=.5*PI-SWEEP
C
C.........  Calculate freestream quantities.
C
      RHOINF=1
```

```
      HINF=1/(GAMMA-1)+.5*MINF**2
      PINF=1/GAMMA
      CINF=1.
C
      UINF=MINF*COS(ALPHA)*COS(YAW)*CINF
      VINF=MINF*            SIN(YAW)*CINF
      WINF=MINF*SIN(ALPHA)*COS(YAW)*CINF
C
C.........  Initialize the Runge-Kutta coefficients.
C
      COEFFRK(1)=.25
      COEFFRK(2)=.33
      COEFFRK(3)=.50
      COEFFRK(4)=1.0
C
C
C
      RETURN
      END
```

```
      SUBROUTINE READRS
C
C$$$$$$$$$$$$$$$$$$$$$$$$$$$$$$$$$$$$$$$$$$$$$$$$$$$$$$$$$$$$$$$$$$$$$
C$$$$$$$$$$$$$$$$$$$$$$$$$$$$$$$$$$$$$$$$$$$$$$$$$$$$$$$$$$$$$$$$$$$$$
C$$$$$                                                           $$$$$
C$$$$$    This subroutine reads the restart file.                $$$$$
C$$$$$                                                           $$$$$
C$$$$$$$$$$$$$$$$$$$$$$$$$$$$$$$$$$$$$$$$$$$$$$$$$$$$$$$$$$$$$$$$$$$$$
C$$$$$$$$$$$$$$$$$$$$$$$$$$$$$$$$$$$$$$$$$$$$$$$$$$$$$$$$$$$$$$$$$$$$$
C
C
      INCLUDE 'LEVIS.INC'
      INTEGER IDUMMY
      REAL XDUMMY
      REAL*8 XXDUMMY
C
C
C
      OPEN(UNIT=3,STATUS='OLD',FORM='UNFORMATTED',FILE='LEVISRS.DAT')
      READ(3) IDUMMY,IDUMMY,IDUMMY
      READ(3) XDUMMY
      READ(3) XDUMMY,XDUMMY,XDUMMY
      DO I=INODBEG,INODEND
         READ(3) XXDUMMY,XXDUMMY
      ENDDO
      DO I=INODBEG,INODEND
         READ(3) RHO(I),RHOU(I),RHOV(I),RHOW(I),RHOE(I)
      ENDDO
      DO I=INODEND+1,NNODESMX
         RHO (I)=RHOINF
         RHOU(I)=RHOINF*UINF
         RHOV(I)=RHOINF*VINF
         RHOW(I)=RHOINF*WINF
         RHOE(I)=RHOINF*HINF-PINF
         P   (I)=PINF
      ENDDO
C
C.......... Calculate pressures.
C
      CALL CALPRS
      CLOSE(UNIT=3)
C
C
C
      RETURN
      END
```

```
      SUBROUTINE SAVRES
C
C$$$$$$$$$$$$$$$$$$$$$$$$$$$$$$$$$$$$$$$$$$$$$$$$$$$$$$$$$$$$$$$$$$$$
C$$$$$$$$$$$$$$$$$$$$$$$$$$$$$$$$$$$$$$$$$$$$$$$$$$$$$$$$$$$$$$$$$$$$
C$$$$$                                                          $$$$$
C$$$$$    This subroutine saves the state vector.               $$$$$
C$$$$$                                                          $$$$$
C$$$$$$$$$$$$$$$$$$$$$$$$$$$$$$$$$$$$$$$$$$$$$$$$$$$$$$$$$$$$$$$$$$$$
C$$$$$$$$$$$$$$$$$$$$$$$$$$$$$$$$$$$$$$$$$$$$$$$$$$$$$$$$$$$$$$$$$$$$
C
      INCLUDE 'LEVIS.INC'
C
C.........  Loop over the nodes
C
      DO INODE=INODBEG,INODEND
C
C.........  Save the state vectors.
C
         RHOO (INODE)=RHO (INODE)
         RHOUO(INODE)=RHOU(INODE)
         RHOVO(INODE)=RHOV(INODE)
         RHOWO(INODE)=RHOW(INODE)
         RHOEO(INODE)=RHOE(INODE)
C
      ENDDO
C
C
C
      RETURN
      END
```

```
      SUBROUTINE SECDIF(VEC,WT,D2VEC)
C
C$$$$$$$$$$$$$$$$$$$$$$$$$$$$$$$$$$$$$$$$$$$$$$$$$$$$$$$$$$$$$$$$$$$$
C$$$$$$$$$$$$$$$$$$$$$$$$$$$$$$$$$$$$$$$$$$$$$$$$$$$$$$$$$$$$$$$$$$$$
C$$$$$                                                          $$$$$
C$$$$$    This subroutine carries out a weighted second difference. $$$$$
C$$$$$    Distributions are carried out in such a way as to avoid   $$$$$
C$$$$$    dependencies.                                         $$$$$
C$$$$$                                                          $$$$$
C$$$$$$$$$$$$$$$$$$$$$$$$$$$$$$$$$$$$$$$$$$$$$$$$$$$$$$$$$$$$$$$$$$$$
C$$$$$$$$$$$$$$$$$$$$$$$$$$$$$$$$$$$$$$$$$$$$$$$$$$$$$$$$$$$$$$$$$$$$
C
      INCLUDE 'LEVIS.INC'
      REAL VEC   (NNODESMX),
     &     WT    (NNODESMX),
     &     D2VEC (NNODESMX),
     &     TEMP1 (NNODESMX),
     &     TEMP2 (NNODESMX),
     &     TEMP3 (NNODESMX),
     &     TEMP4 (NNODESMX)
C
C.........  Statement function for face switches.
C
      MU(WT1,WT2)=MAX(WT1,WT2)
C
C.........  Visit the nodes.  Zero the output vector.
C
      DO INODE=INODBEG,INODEND
         D2VEC(INODE)=0
      ENDDO
C
C.........  Calculate the face fluxes.
C
      DO IFACE=IFA1BEG,IFA1END
         TEMP1(IFACE)=MU(WT(IFA1NOD(1,IFACE)),WT(IFA1NOD(2,IFACE)))*
     &        (VEC(IFA1NOD(2,IFACE))-VEC(IFA1NOD(1,IFACE)))
      ENDDO
      DO IFACE=IFA2BEG,IFA2END
         TEMP2(IFACE)=MU(WT(IFA2NOD(1,IFACE)),WT(IFA2NOD(2,IFACE)))*
     &        (VEC(IFA2NOD(2,IFACE))-VEC(IFA2NOD(1,IFACE)))
      ENDDO
C
C.........  Calculate the cell fluxes.
C
      DO ICELL=ICELBEG,ICELEND
         TEMP3(ICELL)=MU(WT(ICELNOD(1,ICELL)),WT(ICELNOD(3,ICELL)))*
     &        (VEC(ICELNOD(3,ICELL))-VEC(ICELNOD(1,ICELL)))
      ENDDO
```

```
      DO ICELL=ICELBEG,ICELEND
        TEMP4(ICELL)=MU(WT(ICELNOD(2,ICELL)),WT(ICELNOD(4,ICELL)))*
     &              (VEC(ICELNOD(4,ICELL))-VEC(ICELNOD(2,ICELL)))
      ENDDO
C
C.........  Distribute the face fluxes.
C
CVD$  NODEPCHK
      DO IFACE=IFA1BEG,IFA1END
        D2VEC(IFA1NOD(1,IFACE))=D2VEC(IFA1NOD(1,IFACE))+TEMP1(IFACE)
      ENDDO
CVD$  NODEPCHK
      DO IFACE=IFA1BEG,IFA1END
        D2VEC(IFA1NOD(2,IFACE))=D2VEC(IFA1NOD(2,IFACE))-TEMP1(IFACE)
      ENDDO
CVD$  NODEPCHK
      DO IFACE=IFA2BEG,IFA2END
        D2VEC(IFA2NOD(1,IFACE))=D2VEC(IFA2NOD(1,IFACE))+TEMP2(IFACE)
      ENDDO
CVD$  NODEPCHK
      DO IFACE=IFA2BEG,IFA2END
        D2VEC(IFA2NOD(2,IFACE))=D2VEC(IFA2NOD(2,IFACE))-TEMP2(IFACE)
      ENDDO
C
C.........  Distribute the cell fluxes.
C
CVD$  NODEPCHK
      DO ICELL=ICELBEG,ICELEND
        D2VEC(ICELNOD(1,ICELL))=D2VEC(ICELNOD(1,ICELL))+TEMP3(ICELL)
      ENDDO
CVD$  NODEPCHK
      DO ICELL=ICELBEG,ICELEND
        D2VEC(ICELNOD(3,ICELL))=D2VEC(ICELNOD(3,ICELL))-TEMP3(ICELL)
      ENDDO
CVD$  NODEPCHK
      DO ICELL=ICELBEG,ICELEND
        D2VEC(ICELNOD(2,ICELL))=D2VEC(ICELNOD(2,ICELL))+TEMP4(ICELL)
      ENDDO
CVD$  NODEPCHK
      DO ICELL=ICELBEG,ICELEND
        D2VEC(ICELNOD(4,ICELL))=D2VEC(ICELNOD(4,ICELL))-TEMP4(ICELL)
      ENDDO
C
C
C
      RETURN
      END
```

```
      SUBROUTINE SECDNW(VEC,D2VEC)
C
C$$$$$$$$$$$$$$$$$$$$$$$$$$$$$$$$$$$$$$$$$$$$$$$$$$$$$$$$$$$$$$$$$$$$
C$$$$$$$$$$$$$$$$$$$$$$$$$$$$$$$$$$$$$$$$$$$$$$$$$$$$$$$$$$$$$$$$$$$$
C$$$$$                                                          $$$$$
C$$$$$   This subroutine carries out an unweighted second       $$$$$
C$$$$$   difference. The distributions avoids dependencies.     $$$$$
C$$$$$                                                          $$$$$
C$$$$$$$$$$$$$$$$$$$$$$$$$$$$$$$$$$$$$$$$$$$$$$$$$$$$$$$$$$$$$$$$$$$$
C$$$$$$$$$$$$$$$$$$$$$$$$$$$$$$$$$$$$$$$$$$$$$$$$$$$$$$$$$$$$$$$$$$$$
C
      INCLUDE 'LEVIS.INC'
      REAL VEC   (NNODESMX),
     &     D2VEC (NNODESMX),
     &     TEMP  (NNODESMX)
C
C.........  Visit the nodes.  Subtract fix for node.
C
      DO I=INODBEG,INODEND
         D2VEC(I)=-4*NCN(I)*VEC(I)
      ENDDO
C
C.........  Visit the cells.  Calculate the sum of the four nodes.
C
      DO I=ICELBEG,ICELEND
         TEMP(I)=VEC(ICELNOD(1,I))+VEC(ICELNOD(2,I))+
     &           VEC(ICELNOD(3,I))+VEC(ICELNOD(4,I))
      ENDDO
C
C.........  Visit the cells.  Distribute contribution to the sw nodes.
C
CVD$ NODEPCHK
      DO I=ICELBEG,ICELEND
         D2VEC(ICELNOD(1,I))=D2VEC(ICELNOD(1,I))+TEMP(I)
      ENDDO
C
C.........  Visit the cells.  Distribute contribution to the se nodes.
C
CVD$ NODEPCHK
      DO I=ICELBEG,ICELEND
         D2VEC(ICELNOD(2,I))=D2VEC(ICELNOD(2,I))+TEMP(I)
      ENDDO
C
C.........  Visit the cells.  Distribute contribution to the ne nodes.
C
CVD$ NODEPCHK
      DO I=ICELBEG,ICELEND
         D2VEC(ICELNOD(3,I))=D2VEC(ICELNOD(3,I))+TEMP(I)
```

```
      ENDDO
C
C.........  Visit the cells.  Distribute contribution to the nw nodes.
C
CVD$  NODEPCHK
      DO I=ICELBEG,ICELEND
         D2VEC(ICELNOD(4,I))=D2VEC(ICELNOD(4,I))+TEMP(I)
      ENDDO
C
C
C
      RETURN
      END
```

```
      SUBROUTINE SETICS
C
C$$$$$$$$$$$$$$$$$$$$$$$$$$$$$$$$$$$$$$$$$$$$$$$$$$$$$$$$$$$$$$$$$$$$
C$$$$$$$$$$$$$$$$$$$$$$$$$$$$$$$$$$$$$$$$$$$$$$$$$$$$$$$$$$$$$$$$$$$$
C$$$$$                                                          $$$$$
C$$$$$   This subroutine sets freestream initial conditions.    $$$$$
C$$$$$                                                          $$$$$
C$$$$$$$$$$$$$$$$$$$$$$$$$$$$$$$$$$$$$$$$$$$$$$$$$$$$$$$$$$$$$$$$$$$$
C$$$$$$$$$$$$$$$$$$$$$$$$$$$$$$$$$$$$$$$$$$$$$$$$$$$$$$$$$$$$$$$$$$$$
C
      INCLUDE 'LEVIS.INC'
C
C.........  Loop over all the nodes.  Include dummy node.
C
      DO I=INODBEG,NNODESMX
C
C.........  Set the variables to their freestream values.
C
         RHO (I)=RHOINF
         RHOU(I)=RHOINF*UINF
         RHOV(I)=RHOINF*VINF
         RHOW(I)=RHOINF*WINF
         RHOE(I)=RHOINF*HINF-PINF
         P   (I)=PINF
C
      ENDDO
C
C.........  Set the tangency condition at the wall.
C
      DO I=1,NNODWL
         INODE=INODWL(I)
         RHOW(INODE)=0
      ENDDO
C
C
C
      RETURN
      END
```

```
      SUBROUTINE SUMFLX
C
C$$$$$$$$$$$$$$$$$$$$$$$$$$$$$$$$$$$$$$$$$$$$$$$$$$$$$$$$$$$$$$$$$$$$
C$$$$$$$$$$$$$$$$$$$$$$$$$$$$$$$$$$$$$$$$$$$$$$$$$$$$$$$$$$$$$$$$$$$$
C$$$$$                                                          $$$$$
C$$$$$   This subroutine sums the fluxes for each cell.         $$$$$
C$$$$$                                                          $$$$$
C$$$$$$$$$$$$$$$$$$$$$$$$$$$$$$$$$$$$$$$$$$$$$$$$$$$$$$$$$$$$$$$$$$$$
C$$$$$$$$$$$$$$$$$$$$$$$$$$$$$$$$$$$$$$$$$$$$$$$$$$$$$$$$$$$$$$$$$$$$
C
      INCLUDE 'LEVIS.INC'
      REAL SOURCE1(NNODESMX),
     &     SOURCE2(NNODESMX),
     &     SOURCE3(NNODESMX),
     &     SOURCE4(NNODESMX),
     &     SOURCE5(NNODESMX)
C
C.........  Source terms.
C
      DO I=INODBEG,INODEND
C
         SOURCE1(I)=2*RHOU(I)
         SOURCE2(I)=2*RHOU(I)*RHOU(I)/RHO(I)+2*P(I)
         SOURCE3(I)=2*RHOU(I)*RHOV(I)/RHO(I)
         SOURCE4(I)=2*RHOU(I)*RHOW(I)/RHO(I)
         SOURCE5(I)=2*RHOU(I)*RHOE(I)/RHO(I)+2*RHOU(I)*P(I)/RHO(I)
C
      ENDDO
C
C.........  Sum the fluxes.  Visit the cells.
C
      DO ICELL=ICELBEG,ICELEND
C
C.........  Define the faces.
C
         IS=ICELFAC(1,ICELL)
         IE=ICELFAC(2,ICELL)
         IN=ICELFAC(3,ICELL)
         IW=ICELFAC(4,ICELL)
C
C.........  Define the nodes.
C
         ISW=ICELNOD(1,ICELL)
         ISE=ICELNOD(2,ICELL)
         INE=ICELNOD(3,ICELL)
         INW=ICELNOD(4,ICELL)
C
C.........  Sum the face fluxes.
```

```
C
        SFLUX1(ICELL)=FLUX1(1,IS)+FLUX1(2,IE)-FLUX1(1,IN)-FLUX1(2,IW)
        SFLUX2(ICELL)=FLUX2(1,IS)+FLUX2(2,IE)-FLUX2(1,IN)-FLUX2(2,IW)
        SFLUX3(ICELL)=FLUX3(1,IS)+FLUX3(2,IE)-FLUX3(1,IN)-FLUX3(2,IW)
        SFLUX4(ICELL)=FLUX4(1,IS)+FLUX4(2,IE)-FLUX4(1,IN)-FLUX4(2,IW)
        SFLUX5(ICELL)=FLUX5(1,IS)+FLUX5(2,IE)-FLUX5(1,IN)-FLUX5(2,IW)
C
C.......... Add the source terms.
C
        SOURCER=AREA(ICELL)*.25*(SOURCE1(ISW)+SOURCE1(ISE)+
     &                           SOURCE1(INE)+SOURCE1(INW))
        SOURCEU=AREA(ICELL)*.25*(SOURCE2(ISW)+SOURCE2(ISE)+
     &                           SOURCE2(INE)+SOURCE2(INW))
        SOURCEV=AREA(ICELL)*.25*(SOURCE3(ISW)+SOURCE3(ISE)+
     &                           SOURCE3(INE)+SOURCE3(INW))
        SOURCEW=AREA(ICELL)*.25*(SOURCE4(ISW)+SOURCE4(ISE)+
     &                           SOURCE4(INE)+SOURCE4(INW))
        SOURCEE=AREA(ICELL)*.25*(SOURCE5(ISW)+SOURCE5(ISE)+
     &                           SOURCE5(INE)+SOURCE5(INW))
C
        SFLUX1(ICELL)=SFLUX1(ICELL)+SOURCER
        SFLUX2(ICELL)=SFLUX2(ICELL)+SOURCEU
        SFLUX3(ICELL)=SFLUX3(ICELL)+SOURCEV
        SFLUX4(ICELL)=SFLUX4(ICELL)+SOURCEW
        SFLUX5(ICELL)=SFLUX5(ICELL)+SOURCEE
C
      ENDDO
C
C.......... Interface cells.  Add extra face flux.
C
      DO I=1,NCELINT
        ICELL=ICELINT(I)
        IFACE=ICELFACI(1,I)
        ITYPE=ICELFACI(2,I)
        IF(ITYPE.EQ.1) THEN
            SFLUX1(ICELL)=SFLUX1(ICELL)+FLUX1(1,IFACE)
            SFLUX2(ICELL)=SFLUX2(ICELL)+FLUX2(1,IFACE)
            SFLUX3(ICELL)=SFLUX3(ICELL)+FLUX3(1,IFACE)
            SFLUX4(ICELL)=SFLUX4(ICELL)+FLUX4(1,IFACE)
            SFLUX5(ICELL)=SFLUX5(ICELL)+FLUX5(1,IFACE)
        ELSEIF(ITYPE.EQ.2) THEN
            SFLUX1(ICELL)=SFLUX1(ICELL)+FLUX1(2,IFACE)
            SFLUX2(ICELL)=SFLUX2(ICELL)+FLUX2(2,IFACE)
            SFLUX3(ICELL)=SFLUX3(ICELL)+FLUX3(2,IFACE)
            SFLUX4(ICELL)=SFLUX4(ICELL)+FLUX4(2,IFACE)
            SFLUX5(ICELL)=SFLUX5(ICELL)+FLUX5(2,IFACE)
        ELSEIF(ITYPE.EQ.3) THEN
            SFLUX1(ICELL)=SFLUX1(ICELL)-FLUX1(1,IFACE)
```

```
            SFLUX2(ICELL)=SFLUX2(ICELL)-FLUX2(1,IFACE)
            SFLUX3(ICELL)=SFLUX3(ICELL)-FLUX3(1,IFACE)
            SFLUX4(ICELL)=SFLUX4(ICELL)-FLUX4(1,IFACE)
            SFLUX5(ICELL)=SFLUX5(ICELL)-FLUX5(1,IFACE)
        ELSEIF(ITYPE.EQ.4) THEN
            SFLUX1(ICELL)=SFLUX1(ICELL)-FLUX1(2,IFACE)
            SFLUX2(ICELL)=SFLUX2(ICELL)-FLUX2(2,IFACE)
            SFLUX3(ICELL)=SFLUX3(ICELL)-FLUX3(2,IFACE)
            SFLUX4(ICELL)=SFLUX4(ICELL)-FLUX4(2,IFACE)
            SFLUX5(ICELL)=SFLUX5(ICELL)-FLUX5(2,IFACE)
        ENDIF
      ENDDO
C
C
C
      RETURN
      END
```

```
      SUBROUTINE UPDATE
C
C$$$$$$$$$$$$$$$$$$$$$$$$$$$$$$$$$$$$$$$$$$$$$$$$$$$$$$$$$$$$$$$$$$$$
C$$$$$$$$$$$$$$$$$$$$$$$$$$$$$$$$$$$$$$$$$$$$$$$$$$$$$$$$$$$$$$$$$$$$
C$$$$$                                                          $$$$$
C$$$$$    This subroutine updates the state vectors.            $$$$$
C$$$$$                                                          $$$$$
C$$$$$$$$$$$$$$$$$$$$$$$$$$$$$$$$$$$$$$$$$$$$$$$$$$$$$$$$$$$$$$$$$$$$
C$$$$$$$$$$$$$$$$$$$$$$$$$$$$$$$$$$$$$$$$$$$$$$$$$$$$$$$$$$$$$$$$$$$$
C
      INCLUDE 'LEVIS.INC'
C
      DO I=INODBEG,INODEND
C
        RHO (I)=RHOO (I)-COEFFRK(KSTAGE)*
     &              (DT(I)*DR (I)-CFL*2.8*DAMP1(I))
        RHOU(I)=RHOUO(I)-COEFFRK(KSTAGE)*
     &              (DT(I)*DRU(I)-CFL*2.8*DAMP2(I))
        RHOV(I)=RHOVO(I)-COEFFRK(KSTAGE)*
     &              (DT(I)*DRV(I)-CFL*2.8*DAMP3(I))
        RHOW(I)=RHOWO(I)-COEFFRK(KSTAGE)*
     &              (DT(I)*DRW(I)-CFL*2.8*DAMP4(I))
        RHOE(I)=RHOEO(I)-COEFFRK(KSTAGE)*
     &              (DT(I)*DRE(I)-CFL*2.8*DAMP5(I))
      ENDDO
C
C
C
      RETURN
      END
```

```
      PARAMETER(NCELWEMX=64,NCELSNMX=64,NLEMMAX=4,GAMMA=1.4)
      PARAMETER(NNODESMX =4*(NCELWEMX*NCELSNMX+NCELWEMX+NCELSNMX+1))
C
      IMPLICIT REAL (A-H,M,O-Z)
C
C
C
      REAL*8 X1     (NNODESMX),
     &       X2     (NNODESMX),
     &       AREA   (NNODESMX)
C

      REAL RHO      (NNODESMX),
     &     RHOO     (NNODESMX),
     &     RHOU     (NNODESMX),
     &     RHOUO    (NNODESMX),
     &     RHOV     (NNODESMX),
     &     RHOVO    (NNODESMX),
     &     RHOW     (NNODESMX),
     &     RHOWO    (NNODESMX),
     &     RHOE     (NNODESMX),
     &     RHOEO    (NNODESMX),
C
     &     P        (NNODESMX),
     &     DP       (NNODESMX)
C
      REAL DR       (NNODESMX),
     &     DRU      (NNODESMX),
     &     DRV      (NNODESMX),
     &     DRW      (NNODESMX),
     &     DRE      (NNODESMX),
C
     &     DT       (NNODESMX)
C
      REAL FLUX1 (2,NNODESMX),
     &     FLUX2 (2,NNODESMX),
     &     FLUX3 (2,NNODESMX),
     &     FLUX4 (2,NNODESMX),
     &     FLUX5 (2,NNODESMX),
C
     &     SFLUX1 (NNODESMX),
     &     SFLUX2 (NNODESMX),
     &     SFLUX3 (NNODESMX),
     &     SFLUX4 (NNODESMX),
     &     SFLUX5 (NNODESMX)
C
      REAL DAMP1 (NNODESMX),
     &     DAMP2 (NNODESMX),
```

```
     &         DAMP3 (NNODESMX),
     &         DAMP4 (NNODESMX),
     &         DAMP5 (NNODESMX)
C
C
C
      INTEGER    INODNOD (4,NNODESMX),
     &           IFA1NOD (2,NNODESMX),
     &           IFA2NOD (2,NNODESMX),
     &           ICELNOD (4,NNODESMX),
     &           ICELFAC (4,NNODESMX),
C
     &           INODFF  (4*(NCELWEMX+1)),
     &           INODPW  (4*(NCELSNMX+1)),
     &           INODPE  (4*(NCELSNMX+1)),
     &           INODWL  (4*(NCELWEMX+1)),
     &           IFACWL  (4*(NCELWEMX  ))
C
      INTEGER    INODOV  (4*(NCELWEMX+1))
C
      INTEGER    ICELINT(NNODESMX),
     &           ICELFACI(2,NNODESMX),
     &           ICELNODI(NNODESMX),
     &           NEMWEBEG(NLEMMAX),NEMWEEND(NLEMMAX),
     &           NEMSNBEG(NLEMMAX),NEMSNEND(NLEMMAX),
     &           NEMWE(NLEMMAX),NEMSN(NLEMMAX),
     &           NCN(NNODESMX)
C
C
C
      LOGICAL SYM,RESTART
C
C
C
      COMMON X1,X2,AREA,SIGMA,SYM,
     &       RHO,RHOO,RHOU,RHOUO,RHOV,RHOVO,
     &       RHOW,RHOWO,RHOE,RHOEO,P,DP,
     &       DR,DRU,DRV,DRW,DRE,
     &       DT,CFL,ITERS,
     &       NCELWE,NCELSN,
     &       INODBEG,INODEND,
     &       IFA1BEG,IFA1END,IFA2BEG,IFA2END,
     &       ICELBEG,ICELEND,
     &       INODNOD,IFA1NOD,IFA2NOD,ICELNOD,ICELFAC,
     &       NNODFF,INODFF,NNODPW,INODPW,NNODPE,INODPE,NNODWL,INODWL,
     &       NFACWL,IFACWL,NNODOV,INODOV,NCN,
     &       NCELINT,ICELINT,ICELFACI,ICELNODI,
     &       MINF,RHOINF,UINF,VINF,WINF,HINF,PINF,ALPHA,BETA,CONEANG,
```

```
     &      NEMWEBEG,NEMWEEND,NEMSNBEG,NEMSNEND,NEMWE,NEMSN,NLEM,
     &      FLUX1,FLUX2,FLUX3,FLUX4,FLUX5,
     &      DAMP1,DAMP2,DAMP3,DAMP4,DAMP5,
     &      SFLUX1,SFLUX2,SFLUX3,SFLUX4,SFLUX5,
     &      EPS2,EPS4,COEFFRK(4),KSTAGE,
     &      RESTART
```

32	NCELWE	Number of cells running west-east on base grid
32	NCELSN	Number of cells running north-south on base grid
3.0	SIGMA	Grid stretching factor
1.7	MINF	Free-stream Mach number
12.0	ALPHAD	Angle of attack in degrees
0.0	BETAD	Angle of yaw in degrees
75.0	SWEEPD	Leading edge sweep in degrees
1000	ITERS	Number of iterations
1.00	CFL	CFL number (0-1)
0.01	EPS2	Second-difference damping coefficient
0.001	EPS4	Fourth-difference damping coefficient
.FALSE.	RESTART	Logical variable for restart cases
2	NLEM	Number of levels of embedding
1	NEMWEBEG	Starting point for west-east embedding
21	NEMWEEND	Ending point for west-east embedding
1	NEMSNBEG	Starting point for south-north embedding
7	NEMSNEND	Ending point for south-north embedding
4	NEMWEBEG	Starting point for west-east embedding
20	NEMWEEND	Ending point for west-east embedding
1	NEMSNBEG	Starting point for south-north embedding
6	NEMSNEND	Ending point for south-north embedding

List of Symbols

A	cell area
$Æ$	cell aspect ratio
c	speed of sound
C	conical stream-surface
C, D	convective and dissipative operators for multi-stage
C_p, C_v	specific heats at constant pressure and volume
C_p, C_{p_0}	static and total pressure coefficients
d_{ij}	deviatoric stress tensor
D_2, D_4	second- and fourth-difference damping operators
E	mass-specific energy
$\mathbf{F, G, H}$	Cartesian flux vectors
G	amplification factor in stability analysis
$\hat{\mathbf{G}}, \hat{\mathbf{H}}$	conical flux vectors
h_0	total enthalpy
i, j	contravariant coordinates
k	coefficient of thermal conductivity
L	characteristic length scale
$\bar{L}, L$	weighted and unweighted Laplacian operators
M	mass flux through weak solution discontinuity
M, M_r, M_{cf}	total, radial and cross-flow Mach numbers
$\mathbf{n}$	surface normal
n_i	components of surface normal
p, p_0, p_p	static, total and pitot pressures
P, Q	source terms for Poisson equations
Re_{eq}	equivalent Reynolds number
Re_z	Reynolds number based on z
r, θ, f	sperical polar coordinate system
r, θ, z	cylindrical polar coordinate system
S	entropy
t	time
T	static temperature

u, v, w	Cartesian, spherical or cylindrical velocity components
$\bar{u}, \bar{v}, \bar{w}$	conical velocity components
$\hat{u}, \hat{v}, \hat{w}$	scaled velocity components
U	reference free-stream velocity
$\mathbf{U}$	Cartesian state vector
$\hat{\mathbf{U}}$	Cartesian state vector with total enthalpy replacing energy
V	reference swirl velocity
α	angle of attack
α_N	angle of attack normal to leading-edge
α_k	multi-stage coefficients
β	angle of yaw
γ	ratio of specific heats
Γ	circulation
$\hat{\Gamma}$	scaled circulation
δ	flow angularity measured from x axis
$\delta(x)$	Dirac delta function
δ_{ij}	Kronecker delta tensor
δp	weighting function for second difference
Δ	mesh spacing
Δt	time step for multi-stage scheme
ϵ_2, ϵ_4	artificial viscosity coefficients
ϵ_{ijk}	alternate tensor
η, ζ	conical cross-flow coordinates
λ	CFL number
Λ	leading-edge sweep
μ	viscosity coefficient
ρ	density
σ_{ij}	stress tensor
ϕ	conical similarity variable
$\hat{\phi}$	scaled similarity variable
Φ	reference value for vortex core size
ω	vorticity

Subject Index

Alliant FX/8 58
artificial viscosity 28
    and conservation 35, 40
    boundary conditions 35

Beltrami flow 89
boundary conditions
    far-field boundary 13, 31, 32
    Kutta condition 13, 31, 33, 40, 103
    no-flux condition 13, 31, 40
    symmetry plane condition 31, 33
boundary pointers 39
box scheme 113

cell-to-face pointer 39
cell-to-node pointer 39
CFL number 31
classification of vortex flows 1
comparison of computed and measured angularity 83
comparison of numerical and physical pitot pressures 83
compiler directive statements 60
conical assumption 14
    definition of 14
    range of validity 17
conical Euler equations 15
conical self-similarity 13
    definition of 13
    reasons for assuming 7
conical streamlines 17
conservation of energy 11
conservation of mass 10
conservation of momentum 10
contact discontinuity jump relations 103
convergence history 60
core models 5
    Brown vortex 5
    Burgers vortex 101, 104
        scaling 105
        distribution of loss 106
        position of edge 106
        total pressure set by circulation 106
    conical self-similarity model 106
        derivation 107
        core size 113
        loss in core 113
    Guirard and Zeytounian model 5
    Hall vortex 5, 109
    Mangler and Weber vortex 5
    Stewartson and Hall vortex 110
core size definitions 67
cross-flow Mach number definition 53
cross-flow shock above vortex 44, 130, 138
cross-flow shock between vortices 44, 138

cross-flow shock under vortex 2, 44, 79, 121, 138
cross-flow streamlines 46
    critical points of ODE 15
    definition 15, 46
    ODE satisfied by 15

data structure 39
distribution step 30, 32, 33

effects on total pressure loss of
    angle of attack 78
    artificial viscosity level 67
    artificial viscosity form 72
    grid refinement 70, 72
    leading-edge sweep 78
    Mach number 78
embedding interfaces 33
equivalent Reynolds number 89
error in Crocco's relation 89
Euler equations 9
    Cartesian form derivation 10
    conical form derivation 13
    non-dimensionalization of 12
    range of validity 9
face-to-node pointer 39
far-field boundary condition 13, 32
finite-volume formulation 26
flux integration for interface cells 33

grid generation 19
    embedded regions 25
    Joukowski transformation 21
    Poisson equation method 19, 22

Kutta condition 31, 33, 40, 103

Laplacian
    unweighted 28, 35
    weighted 29, 33, 35
limit cycle in streamline integration 47, 138, 166
local conical self-similarity 108
log singularity of Hall's model 110
lossless solutions 95

modified state vector 29
multi-stage scheme 30

no-flux condition 31, 40
node-to-node pointer 39
normal angle of attack 2, 78
normal force coefficients 56
normal Mach number 2, 78

pitot pressure 51, 83
Polhamus suction analogy 4

savings due to embedded regions 61
second-difference operator 28
secondary separation 6
shock capturing 28
shock jump relations 102
shock loss versus core loss 67
spiral structure of feeding sheet 72
sting effects 148, 162
symmetry plane condition 31, 33

temporal discretization 30
total pressure minimum in sheet 103
trapezoidal integration 27

SUBJECT INDEX

vapor screens 121

vectorization 60

vortex capturing 5, 28

vortex flap concept 148

vortex width and height definition 67

weak solution 101

zero mass-flux condition 13

zeroing continuity damping 76

Addresses of the editors of the series "Notes on Numerical Fluid Mechanics":

Prof. Dr. Ernst Heinrich Hirschel (General Editor)
Herzog-Heinrich-Weg 6
D-8011 Zorneding
Federal Republic of Germany

Prof. Dr. Kozo Fujii
High-Speed Aerodynamics Div.
The ISAS
Yoshinodai 3-1-1, Sagamihara
Kanagawa 229
Japan

Prof. Dr. Keith William Morton
Oxford University Computing Laboratory
Numerical Analysis Group
8-11 Keble Road
Oxford OX1 3QD
Great Britain

Prof. Dr. Earll M. Murman
Department of Aeronautics and Astronautics
Massachusetts Institute of Technology (MIT)
Cambridge, MA 02139
USA

Prof. Dr. Maurizio Pandolfi
Dipartimento di Ingegneria Aeronautica e Spaziale
Politecnico di Torino
Corso Duca Degli Abruzzi, 24
I-10129 Torino
Italy

Prof. Dr. Arthur Rizzi
FFA Stockholm
Box 11021
S-16111 Bromma 11
Sweden

Dr. Bernard Roux
Institut de Mécanique des Fluides
Laboratoire Associè au C.R.N.S. LA 03
1, Rue Honnorat
F-13003 Marseille
France